The Geographer as Scientist

The Geographer as Scientist

Essays on the Scope and Nature
of Geography

by

S. W. WOOLDRIDGE, C.B.E., D.Sc., F.G.S.

Fellow of King's College London and
Professor of Geography
in the University of London, King's College

THOMAS NELSON AND SONS LTD

London Edinburgh Paris Melbourne Toronto and New York

THOMAS NELSON AND SONS LTD
Parkside Works Edinburgh 9
36 Park Street London W1
312 Flinders Street Melbourne C1

302–304 Barclays Bank Building
Commissioner and Kruis Streets
Johannesburg

THOMAS NELSON AND SONS (CANADA) LTD
91–93 Wellington Street West   Toronto 1

THOMAS NELSON AND SONS
19 East 47th Street New York 17

SOCIÉTÉ FRANÇAISE D'EDITIONS NELSON
25 rue Henri Barbusse Paris Vᵉ

———

First published 1956

Contents

v

Contents

Text Illustrations

157

Erratum

pp. vii, 174, 177. The captions to Figures 14 and 16 have been inadvertently transposed. They should read:

Fig. 14 Profiles across the margins of the drift-lands in Hertfordshire.

Fig. 16 The form of the Chalk escarpment, north of London.

Text Illustrations

Foreword

'GOOD wine needs no bush' : alike to those who are familiar with Professor Wooldridge's papers in the well-thumbed journals of geography and to those better acquainted with his books, the pages which follow speak eloquently for themselves. His words blow upon us like a strong wind. Like a strong wind they erode—mistaken beliefs, wrong methods, and ignorance. Like a strong wind they gather materials in their course and reshape them into new meaningful forms ; like a strong wind, too, his words, as they buffet, stimulate thought, action—and reaction. There is a feast of reason in this wide-ranging collection for all those who wish to know (and can make the necessary intellectual effort) what stern tasks the geographer confronts, what he is able to do.

This book brings together periodical papers written during twenty-five years of active teaching and research. Most of them were prepared as formal addresses or lectures and delivered at meetings of the Royal Geographical Society, the Institute of British Geographers, the Geographical Association and Section E of the British Association for the Advancement of Science. In their printed form they thus retain many of those distinguishing features of the lecture—vitality, topicality, *obiter dicta*, pungency, and a didactic purpose directed to a particular audience. As a result, and because these contributions come from a critical, original, and richly informed mind, the attentive reader is amply rewarded for his concentration. He can learn much of the craft of geomorphology as Wooldridge has practised it to explain land-forms in England, and of the principles which it applies. Leaving geological time, but no less involved in the problems of our contemporary environment, the reader can contemplate yet other geographical effects of 'Time's winged chariot' in an analysis of the Anglo-Saxon settlement of south-eastern England. By another approach he can concern himself with the methods and difficulties of teaching geography in schools, training colleges, and universities. Again, he is invited to take note that geographical thinking, which tries to comprehend the totality of an environment, has applications to hard problems of our time, such as those which arise in planning Greater London and in the provision of supplies of sand and gravel on an ever-increasing scale. Lastly, and for some readers the picture which they will most prize, is that revealed of a geographer at work, as scientist and as advocate.

There may be still some scientists who so define their calling as to exclude the geographer from their ranks. Certainly geography is not an experimental science. Certainly its subject matter, Man's environments, appears so indecently large and so assuredly complex as to defy understanding by scientific method alone. Yet, as the title of this book asserts and as Wooldridge has vindicated by his own researches, the geographer is or can be a scientist. But, it must be added, for the geographer to be a scientist is not enough. When he leaves strictly measurable phenomena, such as the forms of the land, and is concerned with the relations of social groups to them, he is concerned with materials and considerations familiar to the student of the Humanities and of the Social Sciences.

These pages, therefore, while they have so much directly to convey about the making of the English countryside both by physical and by human agency, and while they put vigorously and unmistakably the writer's concept of geography and of how it should be taught, tell also another story—of how one man becomes a geographer. It is well known how in past centuries geographical knowledge was purveyed by those whose principal activities lay in other fields—discovery and exploration, mathematics, divinity, natural history, geology, history and literature. Even in this century the professional geographer has been commonly recruited from other longer established fields of study. A generation ago S. W. Wooldridge had won his spurs as a geologist and earned his D.Sc. degree by his researches in stratigraphy. It is significant that these researches were made outdoors on the home plot, and that it was the outward witness of the rocks rather than the microscope of the palaeontologist and the mineralogist which engaged his eagerly inquiring mind. His defection from geology and his translation to geography was an act of faith, even if, on the material plane, it reflected the new demand for teachers to undertake, in the University of London, teaching for newly instituted degrees in Geography.

Wooldridge's interest in geography—however much he may have bewailed its earlier formlessness and scientific shortcomings— was from the start central and pregnant : he wanted to understand and explain his own time-moulded country and, as a stratigrapher and field worker—he was never in any sense a 'fossil geologist'—he was booted and spurred for this battle of wits.

The papers collected in this book, mature products of long and concentrated labours of love, measure the victory which Wooldridge has campaigned for and won on behalf of geography. They reveal a range of reading which extends beyond the wide bounds of natural history to include such strange bedfellows as Economics and Holy Scripture. They bear the marks of close study of the ground, of the map, and of the printed text. They leave the reader in no doubt about the message which is being conveyed, for the writing is firm and clear and, when the controversial note is struck, affirmative. A message or a lesson there always is, for the professor is first and foremost a teacher— and how many, colleagues and students alike, have caught something of value from his energy, enthusiasm and inspiration ! At many points in these addresses the cut and thrust of controversy, which is said to enliven the proceedings of geological societies, clarifies the issue as it leaves the reader to judge for himself. To one reader at least the impression is gained, after reading these instructive and combative pages, that a battle for geography has been won, and that commanding heights has been gained and consolidated. A generation of students has been ably disciplined and led ; the paths blazed by W. M. Davis, with whom Wooldridge has temperamental as well as scientific affinities, by Sir H. J. Mackinder, and other earlier leaders have been followed up. The appeal to the young that in the proper study of their habitat they can find and achieve a liberal education has been made and answered. The complaint of geographers that too much has been left to be done by too few need no longer persist : this book indicates clearly what work lies ahead and how to set about it.

The discerning reader will note how both mind and heart have been geared to action in the writing of these papers. He will note how, even as Wooldridge surveys the whole field and shows the relevance of the systematic studies to the main objective of geography—the study of places—he makes clear where his own specialist interests mainly lie. Clearly, among the many mansions of geography, that tenanted by geomorphology offers to him special delight. Clearly, too, for the geographer as scientist this must make a more compelling appeal, even though, with visible evidence inevitably complete, the geomorphologist needs inspiration, as well as measurement, to achieve interpretation. Nor is the reader left in doubt about those academic and pedagogic fundamentals on which Wooldridge continually insists. Such are the indispensability of study in the field and, related to this, his view that geography must begin at home ; the need for careful attention to the physical geography of land-forms and of soils and, arising from this, their significance in explaining patterns of settlement and farming ; the need for a geography which is genetic, not merely descriptive, which turns the geographer back to the past, geological and historical : know the roots and thou shalt know the tree. Yet study of special aspects and of particular problems is not allowed in these pages to obscure study of the whole, although the whole for Wooldridge as for fellow geographers becomes less and less the world and more and more the small (and thus amenable) area selected for study. His main call to students of geography is *circumspice* ; they will not, however, fail to note that the geographical study of places outdoors, at first hand, as distinct from the emotional appreciation of scenery, calls for intellectual application, wide reading and a feeling for country which visibly expresses the unity of man in nature.

W. GORDON EAST

Introduction

THE University teacher of any subject must necessarily and properly require of himself some measure of continuity and direction in his teaching and research which may give coherence to the whole. In no subject is this more insistently necessary than in geography ; not only is it true that students of the subject need such direction and guidance, but one continually faces the fact that the nature, scope, and utility of one's subject is openly or implicitly called in question by colleagues in other subjects. One is concerned to defend the subject as an intellectually respectable discipline and to commend the value of a training in it as a professional qualification. These topics come up for periodical review in inaugural lectures and presidential addresses and similar general statements.

What is attempted in the present work is to gather together a number of such general statements (Essays 1–6) so that they may be considered as a whole, and also to give some examples of actual applications of geographical analysis and investigation, worked examples, as it were, of a procedure readily enough commended in general terms, but only to be fully judged when definitively applied.

British geographers have appeared at times to be little interested in the basic problems of the methodology of their subject ; but that this is not a true judgment of the position is evident from the vigorous discussion which invariably takes place at the characteristic Sunday evening meetings at the annual conference of the Institute of British Geographers. The only British book which deals directly with the matter is *The Making of Geography* [1] by R. E. Dickinson and O. J. R. Howarth, now nearly a quarter of a century old. Both fuller in treatment and more fundamental is an American work by Richard Hartshorne, *On the Nature of Geography*.[2] The great value of this work is that it traces geography from its German origins with full and scholarly attention to the real foundation writings on the subject by Humboldt and Ritter, but it also reports and comments fully

---

[1] O.U.P., 1933   [2] *Ann. Assoc. Amer. Geog.*, 1939

I

upon the later discussions of European and American authors. As a survey of the field of discussion which we here seek to enter it is definitive and indispensable. It is not necessary to agree with all the conclusions reached by Hartshorne to feel heavily in his debt, at least for a thorough clarification of the issues.

As a preface to our present discussion we may briefly review the course of the 'geographical renaissance' in Britain. The starting-point of a long story is found in the report submitted by its Secretary, Sir John Scott Keltie, to the Royal Geographical Society in 1886. At that time geography was taught in schools in nearly all the Continental countries and accorded the same status as other subjects ; it was recognised in nearly all Universities—there were then twelve professorships both in Germany and in Italy and a comparable number in France. As early as 1871 the Society had sought to secure the recognition of this subject at Oxford and Cambridge ; as a sequel to Keltie's report this was achieved at Oxford in 1887 and at Cambridge in 1888. A little before his appointment as Reader at Oxford, H. J. Mackinder delivered his paper, 'On the Scope and Methods of Geography' [1] at the Royal Geographical Society. This is essentially the 'founding document' of British geography. It begins with the question 'What is geography ?', noting that the world, and especially the teaching world, repeats the question 'What is geography ?', with a touch of irony in the tone. The question as he saw it was, 'Can geography be rendered a discipline instead of a mere body of information ?' His point of departure was a recent statement of Sir Frederic Goldsmid at the British Association meeting at Birmingham : 'It is difficult to reconcile the amalgamation of what may be considered "scientific" geography with history.' This point of view Mackinder strongly assailed and rejected. In reading his paper it is necessary to note that contemporary writers were wont to make an antithesis between *physical* geography and *political* geography, using the latter term in the sense in which today one speaks of 'human geography'. The burden of Mackinder's case was that the two were not separate subjects but two stages of our investigation ; he held that 'no *rational* political geography can exist which is

---

[1] Rather surprisingly this paper is not included in the bibliography listed by Hartshorne, op. cit.

not based upon and subsequent to physical geography'. . . . 'Knowledge is after all one, but the extreme specialism of the present-day seems to hide the fact from a certain class of minds.' . . . 'It is the duty of the geographer to build one bridge over an abyss which in the opinion of many is upsetting the equilibrium of our culture'. . . . 'We insist on the teaching and the grasping of geography as a whole.' No modern teacher or student of the subject can fail to find Mackinder's discussion closely and topically relevant. Not all the issues that he defined can be regarded as closed : many of them will concern us in these pages. He defined geography as the science whose main function is 'to trace the interaction of man in society and so much of his environment as varies locally', or alternatively as 'the science which traces the arrangement of things in general on the Earth's surface, considering what relations hold between the distribution of various sets of features on the Earth's surface and what are the causes of their distribution'. Both statements are arguable and did not meet with uniform favour at the time. Yet the success of the initiative of which Mackinder's paper was essentially the opening statement can be judged in the light of the following figures. In 1955 there were over 1,700 students reading for Honours degrees in Geography at British universities, and over 200 post-graduate students. There are at present twenty-nine chairs of geography in the British Isles (1956). The first of these was established at University College, London in 1903. Other important and influential beginnings were made at Liverpool in 1916 (P. M. Roxby) and Aberystwyth 1917 (H. J. Fleure). Full recognition was not achieved at Oxford till 1932 and at Cambridge till 1931. The years since 1945 have seen a very marked expansion with the establishment of no fewer than twelve chairs in British universities and colleges. The growth of the subject has twice reflected what has been called 'the sudden rise of geographical prestige that occurs in wartime'.[1] So large an expansion might suggest that the problems which Mackinder discussed were now settled, but this is far from true. The note of irony is still heard in the question, 'What is geography ?'. Or, as indicated by Professor E. G. R. Taylor, the question

[1] 'Geography in War and Peace', *Report Brit. Assoc. Adv. Sci.* (Dundee, 1947. Pres. Address to Sect. E)

3

'What have they done ?' once asked concerning the fellows of the newly incorporated Royal Society 'is asked today and in the same mocking spirit about geographers'.[1] That is the main reason for bringing together the material of the present volume, and we may now comment briefly on the content of the successive essays and their relation to one another, and to the general questions at issue.

The first essay is primarily concerned to urge that the geographical point of view is as applicable to physical and natural as to social phenomena : to deny, that is, the view that 'the only true geography is human geography'. It calls also for more thoughtful and deliberate development of divisions or aspects of the subject, such as economic and political geography, so that these may take their proper place as specialisms beside those which are more definitely 'scientific' in the narrower or popular sense of that term.

The second essay refers to the change in the nature of the subject which inevitably takes place when it is transferred from the school to the University level and emphasises particularly the virtues of the subject as a corrective to a too-exclusively analytical style of thinking.

The third essay protests at the widespread suppression of the teaching of physical geography at the School level, and notes in this connection the danger implicit in any attempt simply to merge history and geography in a subject entitled 'Social Studies'. Such a subject may well have its place, but as a 'substitute' for geography it appears unbalanced and likely to confirm a tendency to ignore the physical or natural environment. The positive educational value of geography taught in the content of natural history rather than social studies is affirmed.

The following essay treats a subject of great importance and perhaps less disputable, since the opinions here expressed would be widely, though not universally, accepted by British geographers. The theme is the central importance of 'regional geography' regarded as the core and culmination of the subject. It is to be noted that this view of the subject accords with that commonly expressed by educated non-specialist opinion. For the so-called 'man in the street', as for one's fellow teachers, it is probably

[1] *ibid.*

Introduction

true to say that geography is primarily and essentially the 'study of places'. Support from such quarters, while significant and valuable, is not sufficient in itself to settle the point at issue, and attention is directed to the difficulties and imperfections of much regional teaching, which to some extent at least are due to the youth of the subject. The appropriate rôle of contributary specialist studies is also discussed.

Essay the fifth examines in particular some of the problems of the common borderland of geology and geography commonly styled Geomorphology.

The succeeding essays are actual studies in geomorphology ; though they return in places to the more methodological issues (pp. 147-9), they are as a whole designed to illustrate the necessity for an explanatory description of land-forms as an integral part of geographical synthesis. The question, put as tersely as possible, simply is : are these studies properly regarded as geology ?— 'mere geology' as some geographers might be tempted to say ; or do they, as their author would maintain, make some appropriate contribution to regional geography ? Particular attention may be drawn to the course of the argument in the tenth Essay, which begins with geological matter but arrives at regional conclusions applicable in the field of historical geography (pp. 163-70).

This essay serves as a link with the following group which attempt the same form of reasoning, and are concerned in the spirit of the following words of the historian J. R. Green in his *Making of England*.[1] 'The ground itself, where we can read the information it affords, is, whether in the account of the Conquest or in that of the settlement of Britain, the fullest and the most certain of documents. Physical geography has still its part to play in the written record of that human history to which it gives so much of its shape and form.'

This method is applied in the eleventh Essay to elucidate the historical and agricultural geography of the country north and east of London. In the following essay it is more explicitly used as a tool in resolving the problems of Anglo-Saxon settlement. For a fuller treatment of the same theme reference may be made to Chapter III of *The Historical Geography of England before 1800*.[2]

[1] Macmillan (1881), Preface, p. vii
(1,780)

[2] Ed. H. E. Darby (C.U.P., 1936)

In the thirteenth essay the importance of soil conditions in controlling the settlements of the Southeast, both in prehistoric and early historical times, is stressed. It surely must be judged a very significant geographical correlation that early settlement clustered in exactly those areas which remain to this day the chief areas of arable farming. It might appear that all that was required to reach this conclusion was a soil-map. It must therefore be pointed out that no soil maps properly so-called of the area exist. The contention which this group of papers seeks to establish is that in essaying historical studies the geographer can only fully and properly comprehend the physique of the country in question by way of geomorphology.

Exactly the same is true in the field of what we may call 'Applied Geography', i.e. the application of geographical techniques in posing and solving the problems of Town and Country Planning, etc.

The fifteenth essay is, in effect, a summary of the seventeen reports published between 1944–54 by the Advisory Committee on sand and gravel appointed by the Minister of Town and Country Planning. On this Committee the author and his colleague Professor S. H. Beaver served as independent members, under the Chairmanship of Sir Arnold Waters, in association with members of the industry and representatives of other ministries. The results of the inquiry constitute what is perhaps one of the most detailed and successful pieces of planning undertaken in Britain since the Second World War, and in this work the relevance of geographical training in several of its branches became fully manifest. The parent ministry is at present employing some fifty graduates in geography in the work of research and administration. Similar work is also in progress in the Ministry of Agriculture and Fisheries. The fourteenth essay reviews similar problems in the context of the 'Greater London Regional Plan', while the last essay recurs to the problem of land-use in Britain in wider and more general terms.

I should like in conclusion to thank my friend Professor W. G. East for his Foreword which accurately interprets my point of view and also to express my gratitude to the Publishers for their care in seeing a rather difficult manuscript through the Press.

S.W.W.

1

The Geographer as Scientist

THE geographer confronted with the ordeal—or afforded the opportunity—of such an occasion as this has little freedom of choice in his subject. The nature of geography ensures that if he selects some specialist province for treatment he will be speaking a language unknown and distasteful, not only to non-geographers, but to some at least of his geographical colleagues. But the problem of the nature of geography itself concerns not only geographers, but their colleagues in other subjects ; and even if it has been widely discussed, it is still most imperfectly understood. To this problem, therefore, I turn.

The time-honoured scholastic method of tracing the growth of a subject from its presumed origins, is singularly ill-fitted to the case of geography. No doubt the geographic spirit is deeply implanted in the human mind and the historian of the subject may disinter its roots in classical antiquity. Yet ancient geography was generically different from, though of course not unrelated to, its modern representative. The subject suffered a 'sea-change' which began with the 'great age of discovery', and was completed in the era of which Lyell's *Principles of Geology* and Darwin's *Origin of Species* stand as indices. Mother of the natural sciences geography may have been, and she sometimes timidly claims rank, at least as foster parent of the precocious brood of social sciences. But to be compendium of the cosmos was one rôle ; to share and sometimes to dispute the field with a horde of omnivorous offspring is distinctly another. It is this fact that gives such an air of unreality to quotations from classical and medieval writers purporting to foreshadow the modes of thought of modern geography. We need not deny that certain analogies can be perceived. We shall readily commend, in their proper field, studies in the early history of the practical arts of cartography and navigation. But it is a far cry from these to what modern geography claims or attempts, and it is idle to pretend that 'pre-scientific' writers even dimly perceived, let alone clearly formulated, the problems of a geographer in the modern world. Herein lies a main reason for the choice of the title of this lecture. It is in seeming antithesis but in no sense antagonistic to the concept of 'the geographer as humanist', recently treated by Professor Fitzgerald.[1] That the geographer seeks in very large degree the rôle of humanist, is common ground amongst us. Without it most of our prospect and horizon would be gone. Nevertheless, geography, by its very spirit and service, seeks to present its face in two directions. It will conduce neither to clarity nor penetration of vision if, on this occasion, we try to look in two directions at once. If I choose to look science-ward it is because I was trained in science, inherit inescapably its point of view, and can essay to speak its language. And there are words meet to be said in this language to the group of self-regarding and divergent specialisms which, in current usage, pre-empt the name of science. It is from students of these specialisms that much implied criticism of geography comes, and to them an answer is due.

I am not proposing to dispute whether geography is, in the narrow sense, a science or an art ; it is always a barren disputation. In a mere dialectical exercise I should be prepared to plead alternatively that it was neither or both. But I have no doubt that geography can sustain a vital rôle in scientific education. In any case it can be denied the status of science only by narrowing the connotation of this word. A statement, probably apocryphal, but sometimes attributed to Lord Rutherford, avers that science may be divided into physics and stamp-collecting. The implied distinction is presumably as little welcome to biologists and geologists as to geographers. If a mild play on words may be permitted, geography certainly collects stamps, the stamps of natural process and human effort upon the face of our planet. But even if geography be ranged in the stamp-collecting group of sciences, it is accorded by the rest of the group little more than a sedimentary position. To the specialist students of the rock, the plant, the mollusc, insect, or fish, geography seems to abandon the great principle of abstraction by which alone, in their view, science has progressed. It is designedly and obstinately non-

[1] *Nature* **156** (1945), 355

specialist—'everything by turns and nothing long'. Another apocryphal statement cynically defines geography as 'omniology' or 'geocomics', and even if this is a deliberate overstatement of the criticism, it reveals its essential nature. The root misunderstanding here implied is more disastrous for our educational system and for human thought in general than the cleavage, if such there be, between the exact, and the 'natural history' sciences—physics and chemistry *versus* the rest. If physicists and chemists prove by the nature of their calling blind to the relevance of the geographical method they may be forgiven, for admittedly it concerns them little. It is less easy to forgive the biologists, meteorologists, and our nearer relatives the geologists, for the geographical method is relevant within their own fields.

What then is the geographical method? The clue is given by the word 'Zusammenhang', which we find on almost every other page of the writings of the great German founders of modern geography—Humboldt and Ritter, and those of their followers. And for present purposes I would translate 'Zusammenhang' as 'context'. Our aim is to examine rocks, land-forms, soils, plants, as well as human phenomena, in their natural contexts in area, one to another and all together. We need not deny that there are difficulties of method in such study and permissible differences of emphasis in its pursuit. But the reality and necessity of the method and point of view can be denied only by the ignorant or the perverse. The essence of the matter was sufficiently stated by Hettner, forty years ago, when he wrote :

Reality is simultaneously a three-dimensional space which we must examine from three different points of view in order to comprehend the whole ; examination from but one of these points of view alone is one-sided and does not exhaust the whole. From one point of view we see the relations of similar things, from the second the development in time and from the third the arrangement and division in space. Reality as a whole cannot be encompassed entirely in the systematic sciences, sciences defined by the objects they study, as many students still think. Other writers have effectively based the justification for the historical sciences on the necessity of a special conception of development in time. But this leaves science still two-dimensional ; we do not perceive it completely unless we consider it also from the third point of view, the division and arrangement in space.[1]

[1] *Die Geographie, ihre Geschichte, ihr Wesen und ihre Methoden* (Breslau, 1927), p. 114, *et seq*, trans. R. Hartshorne, *Ann. Assoc. Amer. Geog.* 29 (1939), 140

To which we might add, as pendant, the statement of Kraft [1] : 'Stones, plants, animals, and man, in themselves objects of their own sciences, constitute objects in the sphere of geography in so far as they are of importance for, or characteristic of, the nature of the earth's surface.'

I am well aware that statements in such semi-philosophical language make small appeal to pragmatic British ears, yet the essential truth of Hettner's thesis is not to be gainsaid. No colleague of mine has ever to my knowledge sought to deny this thesis ; indeed, their behaviour suggests that they are in no way aware that there is a thesis either to affirm or deny. The most that one customarily gets is an easy-going tolerance—the assumption that since geography departments exist, there is presumably something worth studying inside them, though to all appearances it is a pretentious, shallow, formless, and undisciplined subject. It is common knowledge that the Faculty of Science of this University has been engaged in reshaping the syllabuses for our Special Science degrees. One admirable feature of the new scheme is the admission of ancillary subjects as integral parts of Special Honours courses. Assuming roughly two such subjects in each case, the main specialism will commonly name one ancillary as obligatory or highly desirable ; mathematics for physics, physics for chemistry, chemistry for botany and so on, and one or more additional subjects to be selected at will from a list. It is natural that no subject asks for geography as a 'first' ancillary ; it is deplorable that few of them, none I think except geology, admit it as even a permissible second choice. The geologists did so only very reluctantly at the instance of a few individuals, who were concerned, in this instance, with the well-being of geology, not of geography. It would appear, indeed, that scientists as a class are not at all keenly aware of Hettner's third point of view, the study of the arrangement and division of phenomena in space. Their possible reply would be that they feel quite capable of dealing themselves with the geographical aspects of their own subjects. How far this claim can be substantiated in fact we shall examine below. But even if it were fully conceded, it remains true, in my judgment, that there is a

[1] 'Die Geographie als Wissenschaft' *Enzykl. der Erdk.* (Vienna, 1929), p. 4, trans. as above, p. 143

common tendency to underrate the geographical method, to mistake narrowness, and, in a sense, depth of field for intensity of thought.

The point at issue is well exemplified within the boundaries of the science of geology itself. Let us note first that geology does, or at least geology can, embrace within its own confines the three points of view distinguished by Hettner ; it deals in an analytical, classificatory way with special classes of objects, minerals, rocks, and fossils, but it is also both an historical and a geographical study, in Hettner's sense. For this reason I believe that a training in geology affords one of the best possible educations in the scientific curriculum. Yet at present it enjoys no adequately large or favoured place in university studies, and attracts few students. For this there may be many reasons ; as Professor H. H. Read has recently observed, there has been much wailing at the present plight of geology, but few effective remedies proposed. May I indicate one of the reasons and one of the remedies ? A geologist, as Sir John Adams was wont to point out in his lectures on education, is concerned in two very different sorts of work. Observe him poring over crystal fragments or microscopic fossils on a laboratory bench and he appears to be working hard. Anon you find him leaning on a five-barred gate regarding the view ; he may even be smoking. Yet he is working just as hard. But there has been a strong tendency to magnify the importance and prestige of laboratory work at the expense of the non-laboratory aspects. The great geologists of the later nineteenth century were general practitioners of wide and catholic view, but weak sometimes in the specialist contributory disciplines of mineral and fossil. These they perforce left to their assistants, who in time succeeded them. These latter in some cases—I must emphatically state, not in all—then proceeded to put the cart before the horse. They were rather bored by what may be broadly called the historical and geographical aspects of their subject and damned work in these fields with faint praise. I was not unaware of this attitude as a young research worker. I was engaged in work, geographical at least in the sense that it was palaeogeographical. I was interested then, as I am now, in the great trench of the lower Lea valley, which in geological time past as in the geographical present has divided the London Basin, north of the Thames, into two significantly contrasted regions. I sought to delimit the marine from the freshwater facies of the Lower Bagshot sands and discovered that the line of separation coincided with the Lea valley. This was only a detail, in a long and complex story, but as I stood then looking over the great Lea trench, I took pleasure in the spatial relationships of things and noted how an old-established structural line controlled the lineaments of a vanished geological landscape and controls not less present scenery and human economy. But some of my friends were of other mind. 'After all,' said they, 'you are only drawing a line on a map. We, on the other hand, are well on the way to becoming masters of a fossil group. Day by day and in every way we know more and more about less and less ; but that is the way to international fame and reputation.' Let nothing I say conceal my realisation of the value and necessity of laboratory study. To the petrologists at least I can say 'et ego in arcadia'. But geology is in danger of surrendering its cultural heritage for a mess of indigestible pottage. With all deference to my distinguished colleague Professor Bernal, it will be an ill day for the subject when it becomes a mere appendix to crystal physics, nor do I see the faintest reason to suppose that if it moved in this direction it would, in his words, 'attract a better class of mind than at present'.[1]

This brief, and as some may think impertinent, review of the internal problems of another subject, is by no means irrelevant to our main theme, for the point of view which shows signs of atrophy in geology, is even more conspicuously absent in other parts of the scientific field, both natural and social. Whatever be the value and validity of the geographical point of view in geology, I affirm its necessity and importance as a general method. The synoptic, comparative view of phenomena, the carefully cultivated habit of seeing woods rather than trees, is one means of satisfying man's intellectual curiosity, and his æsthetic urge to descry order and regularity in the world about him. And if his aspiration be to interpret what he sees, to distil its significance, then the geographical method is an indispensable road to truth, different from, but supplementary to that pursued by the analytical sciences.

Let me speak for a moment in parable. Suppose the Master

[1] J. D. Bernal, *The Social Function of Science* (Routledge, 1929), p. 255

of this College together with my colleagues in botany, geology, and myself were cast away on a large, well-found, but unknown and uninhabited, island. Granting us all a Crusoe-like modicum of equipment, my colleagues might beguile the tedium of waiting for rescue, by collecting and examining, say, isopods, fungi, and detrital minerals respectively. And what of my own rôle ? I should have an island for study—an area of the earth's surface. Condemned by my craft to study things as wholes and to observe the great principle of natural context, I should be courteously insistent that the whole was more than the aggregate of isopods plus fungi plus detrital minerals. I would attempt at least to explore the island and map it by plane-table, so that its surface features and the resultant and associated habitat factors might be brought into general review. And if my efforts stopped short of full geographical synthesis they would get at least as far as such work has, in fact, gone over much of the inhabited globe.

It is, of course, no part of my submission that the rôle of geography is simply to assist specialist researches in other fields. My parable illustrates rather the relativity of knowledge and the need for division of labour. It is a rarely gifted mind that can scrutinise both the map and the microscopic field, alternately or both at once, and resolve the mental images stereoscopically as it were. I deliberately simplified the problem by divesting the island of human inhabitants, other than the academic castaways. I have thus deprived the geographer's field of more than half its interest. But within even this narrowed field the principle which informs the spirit and service of geography would find expression, and if physical geography and biogeography had not previously been invented, the Master would have to act as he often does and appoint a sub-committee to invent them.

May I now direct your attention rather more closely to certain special aspects or branches of physical and biogeography. I do not propose to say much of geomorphology, despite my great and abiding interest in it. A man on his special subject easily becomes a bore of the same genus as he who has gone round Lytham St Anne's in two under bogey and wants to tell you how it was done. I permit myself two points only. The first is to announce the simple and obvious truth that geomorphology is the study of land forms. It is not, as the practice of certain geographers seems to imply, a name for the whole geological content of geography, nor as certain geologists appear to assume, a convenient designation for general elementary geology. As regards the first delusion, I was once taken severely to task by a colleague for asking what was in his view a geomorphological question, by requiring a geographical comparison of the several types of Chalk area in Britain. The point is, of course, that they are virtually identical in geomorphology, but differ in situation, soil, plant cover, and human utilisation. But I had mentioned the fatal word 'Chalk' so of course it was geomorphology ! As regards the second fallacy, it is of course true that you cannot teach geographers geomorphology in their third year if they have acquired no earlier 'physical basis' to their subject, but that does not justify you in tacking the name geomorphology on to elementary notions of this planet, its constitution, and the nature of its crust, which they should have learned long before. It is a matter of amazement and distress to me that there are still schools of geography in Britain which, despite lip service to physical geography, devote the most perfunctory attention to geomorphology. If geography, as we now are largely agreed, deals primarily with the differentiation of the surface of the earth, geomorphology must continue to be, as it has always been in Germany and was once in America, an essential part of a geographer's training, though of course only a part.

To pass to my second point which concerns the main theme of this lecture. I am not greatly concerned as to the sharing of geomorphology between geographers and geologists. In this country geologists have to some extent, though by no means wholly, rather contemptuously relegated it to geographers. In its home of origin, the United States, it claims rank as a separate subject. But if that be its real rôle, it remains true that there is still possible a distinctively geographical approach to it. It is this question which to my mind underlay the divergence between W. M. Davis and Walther Penck, recently discussed at length by the Association of American Geographers.[1] The approach of Davis was essentially that of a geographer, of Penck that of a physical or dynamical geologist. Both views are necessary and I do not

[1] 'Walter Penck's Contribution to Geomorphology. A symposium' *Ann. Assoc. Amer. Geog.* 30 (1940), 219

doubt that some geomorphologists have been unduly neglectful of processes—the field of physical geology. But can we deny that Davis threw much light on earth sculpture by studying land-forms *en suite*—in association or mutual context ? It is because human patterns are so delicately related to surface relief, that the geographer will never be able to divorce his study from geo-morphology, properly so-called, and his patient examination of the Earth's external face will not only pose but help to solve problems in the pure study of land-sculpture.

An entirely analogous divergence of aims and method con-cerns the relations of meteorology and climatology. Part of the field of meteorology has proved susceptible of mathematical treatment, and such part or parts are distinguished as physical and dynamical meteorology. The portion so segregated is a relatively small part of the field of reality compassed by our present knowledge of the atmosphere. Synoptic meteorology falls outside it and I venture to think it will do so for a long time to come, for I cannot conceive that formal solutions to equations with so many variables are at all imminent. You cannot rigidly calculate, but only helplessly observe, the rate of progress of a front. For some inscrutable reason, this fact is held to derogate from the scientific stature of weather forecasters, despite their increasing skill in inductive and comparative techniques. I resent this brand of scientific snobbery, this discrimination against natural phenomena, because they prove unamenable to mathe-matics at about Higher Certificate, or indeed any other level, and I seem to scent once again the tendency to underrate an essentially geographical craft. But worse is to follow, Climatology in the words of the authors of a recent textbook [1] is : 'obviously closely related to meteorology and also to geography, since the atmosphere is strongly influenced by the physical nature of its Earth's surface'. 'Moreover,' they continue, 'the simplest method of arranging a description of the various terrestrial climates is to enumerate the climatic conditions according to geographical units'. Both statements are of course unexceptionably true, if not ludicrously obvious. But to imply that geography has no more concern with, and no more contribution to make to climatology than this is to misuse the word almost as radically

---

[1] B. Haurwitz and J. M. Austin, *Climatology* (McGraw, 1944), p. 1

as when you ask a friend if he knows the geography of a hotel you are visiting.

I am not in the least inclined to waste the time of this audience by a ponderous explanation of what climatology really is. Everyone seems to be aware of its nature except certain meteorologists, who, by accident of birth or training, are incapacitated for distinguishing between wood and trees. 'Geography,' it has been said, 'is a social science in so far as it concerns itself with the expression of man in area.' [1] Similarly we may say that geography is a natural science in so far as it concerns itself with the expression of natural phenomena in area—in this case atmospheric phenomena. Statistical or distributional climatology will continue to exist independent of the vagaries of air-mass theory. Some of my younger meteorological friends returning to the field of geography from Service meteorology are assuming a rather preoccupied mien and promising to re-cast the whole field of climatology in a synoptic or dynamical guise. We shall watch their efforts with interest and sympathy. But I would bid them never to forget that a frost-immunity map and, say, an occlusion are incommensurable concepts. We are interested, but only mildly interested, in their fronts which fail to develop normally owing to accidents of inclination, or highly inferential upper air conditions. We are much more interested in what, to use their own new-found language, we may term positive values of the integral θ-42, with respect to time, where θ represents the actual temperature in °F.—or to translate into ordinary English—in accumulated temperature above the basic limit critical for plant growth. [2]

I should desire also to record my opinion that climatology will never be handled in a live and progressive manner, whether in the Air Ministry or outside it, except by men trained in geography, whatever other training they may need. And I cannot fail to note that, despite the almost contemptuous reception

[1] C. O. Sauer and J. B. Leighly, *An Introduction to Geography, I—Elements* (Edwards Bros. Inc., 6th ed., 1922), p. 7

[2] The important data summarised by Brunt (*Nature* 155 (1945), 559) in respect of the physiological reaction to climate of man as an individual animal, represent a part, though a small one, of the field of climatology. The commercial and cultural adaptations to climate of human groups is the prime concern of the geographer, and the classification of climates which ignores these aspects has but a limited value for him.

accorded to certain of my colleagues and students volunteering for meteorological work in the early days of the war, some of them appear to have been capable of giving satisfactory service, even in research branches, because of and not in spite of their geographical training. I could wish that more adequate acknowledgment of this fact had been forthcoming. It is not of course that any informed person doubts the relevance of mathematical training. But more, indeed much more, than mathematics is required.

Let me turn briefly now to the field of plant ecology. In the opening lines of his *Types of British Vegetation* published over thirty years ago, Tansley quoted Drude of Dresden as stating that plants could be studied from three points of view and three only, the physiological, the phylogenetic, and the geographical. This three-fold division is philosophically closely accordant with that of Hettner applied to the whole field of knowledge. In the earlier stages of British work, the geographical aspect—vegetation survey—was well represented by the labours of the brothers Smith and many others. Tansley's other great work on British vegetation[1] is a magnificent register of the progress achieved in later years. But, in certain quarters, the importance and success of work in plant physiology seems to have encouraged the notion that the line of progress by means of survey was 'worked out' and in any case a rather inferior approach. I find the attitude clearly crystallised by Leach[2] who refers to the 'more or less approximately accurate estimate of the more obvious environmental conditions afforded by field work', adding that : 'this somewhat superficial conception of the subject, while excusable in the early days, is quite out of keeping with modern progressive tendencies in biological science'. And again he remarks, 'It should be needless to point out . . . the futility of making merely superficial examination of various examples of a given type of vegetation as they occur in various places in the hope that the reasons for the existence of such a type will eventually reveal themselves'. This is an open dismissal of the geographical method in favour of laboratory investigations. It is needless to commend the latter, but very needful to deplore the threatened

[1] A. G. Tansley, *The British Islands and their Vegetation* (C.U.P., 1939)
[2] W. Leach, *Plant Ecology* (Methuen, 1933), pp. 1, 4

17

supersession of Drude's third approach to the study of plants. If survey is 'quite out of keeping with modern progressive tendencies in biological science' so much the worse for biological science. It has been suggested to me that if there is any remaining need for ecological survey, it must be left to geographers to do it. They will certainly do it, not because they are minded to spend their efforts on unwanted parts of other men's fields, but because natural vegetation is a vital part of human environment and close-linked to land utilisation. But let it be remembered that if, to take an example, they follow Stapledon's suggestion and map the upper limit of bracken on British hillsides, they will be throwing light not only on the land of Britain and its present and future uses, but also on bracken and its associates.

So far I have been dealing almost exclusively with the field of physical or at least of non-human geography ; and in a sense I have been speaking not so much to geographers as for geographers. Geography makes yet other and greater claims, and I have no doubt that it is in the field of human phenomena that it has its greatest opportunity, even if it there faces its greatest difficulties. I would predicate only one thing—that any attempt to take the ge- out of geography and to convert it into what, for want of a better word, I will call socio-anthropography, will doom our subject without hope of appeal, to a further generation in the wilderness. I have not, however, the slightest doubt that the method of study by spatial context is as necessary and as applicable in reviewing human phenomena, as in the more narrowly natural field.

We may admit that serious difficulties have always confronted the attempt to subsume human phenomena under geographical categories. The objection, tacit or explicit, has always been that the mind and spirit of man, not to mention the lamentable and frequent absence of either mind or spirit in man, enters as an incommensurable element and disrupts the chain of causation. Are we committed to include within our study man as everyman —man the martyr, the hero, the thinker, man the merciful, man the son of God, as some believe, a little lower than the angels, and, at the other pole, man behaving with the stupidity and savagery of a cornered rat, or with the leering, self-regarding shrewdness of the defaulting lawyer, or the so-called individualist smart-aleck of the market place. If so, we have plainly bitten off more than we can scientifically chew. Manifestly, geography to save its sanity must make far lesser claims than this. But the answer to this seeming problem is perfectly simple. If, as most of us agree, the central preoccupation of geography is with the differentiation of the Earth's surface, how can it be denied that man is one of the chief differentiating agents, worthy at least to rank with tectonic forces, erosional processes, climatic conditions, and the rest. It is with satisfaction that one notes the curiously similar statements, made quite independently by geographers of significantly different outlook. 'Geography,' say Sauer and Leighly in a passage from which I have already quoted, 'has never been a science of man but the science of the land, of the Earth's surface. It is limited, therefore, on the social side to the attempt to secure valid judgments regarding areal expression of culture.' And in like fashion Vidal de la Blache [1] writes, 'Geography is the science of places, not that of men ; it is interested in the events of history in so far as these bring to work and to light, in the countries where they take place, qualities and potentialities which without them would remain latent.'

Here indeed is a salutary limitation of the field, which yet leaves geography with a quite indispensable central rôle in both scientific and human studies. It is a valid criticism not only of our educational system, but of our whole latter-day world view that we have permitted knowledge in the words of Fleure [2] 'to suffer a deep fission into two incomplete and inadequate parts, the physical and the historic'. It was the late President of this College, the former Archbishop of Canterbury [3] who defined a modern university as 'a place where a multitude of studies are conducted with no relationship between them except simultaneity and juxtaposition'. I would certainly claim that the latter-day rise of geography is one of the few signs of dawning sanity in an educational wilderness of atomistic specialisation, a wilderness which cannot be held wholly unaccountable for the present plight of our species.

So far all geographers will be with me, but they will know

[1] *Ann. de Géog.* **22** (1913), 299
[2] *An Introduction to Geography* (Benn, 1929)
[3] In a sermon before the University of Oxford, quoted by A. S. Nash, *The University and the Modern World* (Macmillan, 1945), p. 183

what lions lurk in the path and how radical is the difference between announcing a project and carrying out a programme. It is no longer a question of defining the aim of geography, but of seeking methods by which the aim may be carried out. This problem of geographical methodology has been discussed almost *ad nauseam* during the last twenty-five years, though British writers have played relatively little part in the fray. Hartshorne,[1] to whose weighty and compendious monograph of 1939 we owe admiration and gratitude, has well remarked that a survey of the literature might make it appear that 'geography was founded by a group of American scholars at the beginning of the twentieth century'. As it now seems, many of us have been guilty of producing from our own hat, with the fatuous complacency of a third-rate conjuror, ideas purporting to be new, but which are actually a half-century and more old. Our destestation of the latter-day aberrations of German thought is no excuse for our common failure to study and to ponder the works of the real founders of our subject, Humboldt and Ritter, and our debt to Hartshorne is the measure of our culpable ignorance of the literature of our own subject.

If, however, we seem to detect and wish further to encourage a certain convergence of aim and method, I have small patience with a certain heresy-hunting proclivity manifest amongst us. If rumour has spoken truly, it has been thought proper by sincere well-wishers of our subject, to condemn the work and damage the prospects of some of the ablest of its younger exponents, by averring that they are 'not really geographers'. I am persuaded that this misguided and uncharitable attitude must cease. I find myself cordially at one with Gradmann when he wrote, 'Is it really completely unavoidable that our science should be exposed naked before all the world, in that every new year a new reformer arises who condemns the whole previous works to the foundations and feels himself called upon to erect something entirely new out of ruins ?'[2]

I too have played this game. May I say to some of my friends how well I remember our discussions in bar-parlours and beside the silent pool of coffee in London tea-shops. I do not

[1] 'The Nature of Geography' *Ann. Assoc. Amer. Geog.* 29 (1939)
[2] *Geog. Ztschr.* 34 (1928), 552, trans. by R. Hartshorne, *op. cit.*, p. 27

regret those discussions ; they were infinitely repaying, but there hung over them like a cloud a certain impatience of precise methodological formulation on the part of our seniors. I remember how we slated the latter for their seeming lethargy of mind and their lofty disregard of problems which convulsed us. 'When I use a word,' said Humpty-Dumpty (the Professors), 'it means just what I choose it shall mean, neither more nor less.' ' The question is,' said Alice (ourselves), 'whether you can make words mean so many different things.' And now the penalties and prominences of seniority have descended on me. Already I see the force of the position which some senior geographers have characteristically taken up. I know perfectly well what I mean by geography and what I believe its spirit and service to be ; I am, with Professor Griffith Taylor [1] and I hope others, a 'geocrat', I shall naturally and inevitably seek to recommend what appear to me to be the advantages of this point of view to my colleagues and students. But nothing in my own experience inclines me to constrain colleagues or students to a rigid uniformity of outlook. And if other University Schools prefer different emphases, though I may nurture in my secret heart the hope that it is *they* who are deluded, I am well aware that they will naturally reciprocate the feeling and I shall be led to ponder the virtues of tolerance.

Methodological discussion of course has its value, and the fires of discussion are productive of light as well as heat. And lest it should seem that I am evading such fires, I should wish to cast, in conclusion, some few mild and entirely friendly apples of discord among my colleagues and students. The emergence of the 'landscape' concept or method, however defined or qualified, is, I judge, a signal advance in geographical thinking. To study, first at least, the imprint of Man on Land and not the effect of Land upon Man, is wise from many points of view. To make too great and exclusive a fetish of the so-called 'cultural landscape', as such, would be indeed to invite a repetition of the crushing rebuke of my predecessor in this Chair, that we had 'invented a jargon and proclaimed a philosophy'. But few serious students of the subject could fail to acclaim the great clarification of technique and aim which springs from this particular emphasis,

[1] 'Correlations and Culture' *Report Brit. Assoc. Adv. Sci.* (Cambridge, 1938, Pres. Address to Sect. E)

which if not new, is at least increasingly and significantly popular.

There has been, however, a tendency unduly to belittle the study of the effect of Land on Man—'environmentalism' as it has been called—and to decry what was once termed 'the geographical factor'. One has heard sneers concerning the supposed 'facile generalisations of anthropo-geography' from other social scientists, who, one suspects, have read little later than Ratzel, if, indeed, they have really read him.

There is no need to deny the patent limitations and possible errors of the environmental approach to geography. Geography is not and never will be competent to handle more than a part of the field of study so indicated. It is not a synonym for 'human ecology'. If it were, how would it deal distinctively with both synecology and autecology? How would it treat environment in an urban slum, and how here delimit its field from that of sociology? And are geographers knowledgeably competent in the field of integrative psychology as treated by Marston? Manifestly not. Yet the botanist J. W. Bews finds it necessary to treat of Marston's work and much other material entirely non-geographical under the heading of 'Human Ecology'.[1]

But all this leaves the fact quite unaltered, that there is a 'geographical factor'. To object, as some have done, that to study influences is to postulate them in advance is merely, I feel, to sport dialectically in a cloud-cuckoo land of sheer unrealities. In the same way Febvre's delicate *Wortspielerei* at the expense of 'influences' in geography seems widely to miss the mark.[2] There is no need to postulate the influences in advance in order to study them. They are given to simple observation and reflection in any inhabited area whatsoever. They are, it is true, legible to the geographer in terms of landscape, not in terms of the nature of individual and communal man. Concerning the latter, we may well protest our continuing ignorance, and geography at best and latest will have only part of the answer. Meanwhile, geography is not accountable for the delusions of the poet Corneille, of the 'Esprit des Lois' and of Henry Buckle, nor the group of radically unscientific writers on 'national

---

[1] *Human Ecology* (O.U.P., 1935)
[2] L. Febvre, *A Geographical Introduction to History* (Kegan Paul, 1925)

character'.[1] But the flight from determinism, however desirable in some of its aspects, bids fair to render us ridiculous. It is hardly too much to say that on Tuesdays and Thursdays we are wont to declare our Simon-pure freedom from any taint of the influence of influences, while on Mondays, Wednesdays, and Fridays, in our research and teaching, we 'describe and try to account for' various human phenomena in terms of their physical background. No doubt it is better to speak of correlations than of influences. No doubt we shall be wise, in Whitehead's words, 'to seek simplicity and distrust it'. But we always have gone and we always shall, in practice, seek to go further than noting the mere simultaneity and juxtaposition of physical and human phenomena, and shall learn much in doing so. The philosophical pre-suppositions of such a question as 'Describe and account for the distribution of population in Scandinavia' are no more to be called in doubt within the tacit universe of discourse, than if we substitute for the last six words 'the foreign policy of Bismarck', or 'the chemical characteristics of esters'. We may grant that physical geography is not the chief or the sole determinant of the characters of human societies, but we affirm that human societies react even to the minutiae of physical geography, in the working out of real patterns of occupancy and economy, and to this extent they are significantly influenced by them. Further, it is such aspects and relations of human societies which it is the prime function of the geographer to elucidate.

My brief and necessarily inadequate reference to some of the internal problems and disputes of human geography may seem to have carried us far from the theme of the geographer as scientist. Such, in fact, is far from the case. That the geographer indeed needs to remain scientific in the best sense, when dealing with the human portion of his field, is obvious enough, but to state such a pious aspiration illuminates the problem but little. Let us say rather that geography shares a perimeter not only with the natural but with the social sciences, and has contacts and borderlands along this frontier with Economics, Sociology, and the other subjects of the group. To them it acts as a kind of physical interpreter, though this does not exhaust its rôle.

The geographer can at least envisage fields of study such as

---

[1] cf. Febvre *op. cit.*

23

political geography and economic geography, properly so-called. I say he can envisage these fields, but he has not in my judgment done much yet to delimit or cultivate them. Let us make an effort to avoid a common confusion here. No-one denies, of course, that the economic and political characteristics and relationships of man as expressed in landscape are part of the central subject matter of geography. They are not political or economic geography but just geography. My plea is that geographers have not yet developed these special aspects so that they can rank as co-ordinate with geomorphology, plant geography, and the like. It is easy enough to tack any number of desired adjectival qualifiers on to the word geography. It is less easy, but perfectly possible, to collect a list of topics which may, for a particular session, be deemed to constitute a course in such 'qualified' geography. But it is very difficult to demonstrate that such fields have any adequate or ordered content as yet, as special borderland subjects. The Board of Studies in Geography in London University has wrestled long and earnestly for four years in seeking to reshape the Honours Geography degree. I should be entirely lacking in frankness if I did not affirm my belief that there is an unresolved confusion behind much of our thinking. Those of my colleagues who are primarily 'humanists'—I use the term in the most affectionate and respectful sense—have gone through life for years with a double-headed penny in their pockets, which enables them to teach and ask questions upon identical subject matter under the heading of both regional geography and of the various qualified geographies we have noted. This seems to me, I must confess, a deplorable and indefensible state of affairs. It can be terminated only by the deliberate and conscious development of the specialist aspects of human geography. To the geographers, and particularly the younger geographers, seeking such development, I would say you must publish much more freely than hitherto in these fields —publish and if necessary be damned. But it is fair to add that the faults of under-development in these specialist human aspects and a certain resulting lop-sidedness in geography in general, reflect also the retarded development of the social sciences themselves. Our own University loses as well as gains by the virtual segregation of such studies within one great institution [1]—

[1] The London School of Economics

24

one I am happy to add, with which I have had long and friendly relations. I am compelled at least to say that the authorities of this college owe it to the Department of Geography to provide at least some teaching in economics. It will require economists and geographers to build an economic geography worthy of the name, and the same principle will be found to have application throughout the human frontiers of geography. Without the support of such subjects, geography, and I think also history, while affording each other much mutual support, are condemned to the rôle of maids of all work and are tempted also to unprofitable trespassing in the fields of others.

As I draw to a close, which many of you I feel must be anxiously awaiting, I reflect that I have only restated tonight, with all the limitations inherent in my own training and outlook, some small part of the general thesis which formed the basis of inaugural addresses by Richthofen at Leipzig in 1885, and by Chisholm at Edinburgh in 1908, and other geographers on uncounted occasions. To my non-geographical colleagues I would say, in summary, that geography still stands where it did, seeking to maintain its distinctive point of view ; it can do no other. With its new found strength it is little minded to be driven into diffident retirement again, by criticisms, often ill-founded, proceeding from a culture, dominantly scientific in the narrower sense, and I venture to think, in the light of the condition of the world, in many respects bankrupt.

To my own students and others, I would say that it is no mean, facile, or superficial picture of the world of reality which geographers will seek to present to you. The subject interested you initially ; you will find that it will grow upon you powerfully. Sometimes you will be tempted to shy at the difficulty of the mental task presented and to envy the blinkered cosiness of narrower and more readily definite investigations in other subjects. But you will learn both in your own work, and in reading that of the great geographers of the past, that width of view is not to be equated with shallowness of mind. The subject will give you mental exercise, as severe and stimulating as you could desire, but it will give you also a point of view of high value in the present world of men, distracted and gravely threatened by the patent limitations of their customary modes of thinking.

Geographical Science in Education

IT HAS taken more than half a century of effort to secure a reasonable measure of recognition for geography in this country and the battle is not yet wholly won. When in years to come the struggle is surveyed in retrospect, certain curious and significant features will become clear. The living idea for which geography stands will be seen to have emerged as a sign of the times. The First World War had come and gone before there was any major advance in the Universities. The next two decades brought probationary recognition for the subject at one University after another in steady succession, followed in some cases by full recognition and by the creation of a chair in the subject. The period of the Second World War saw further progress; today there is only one University in England and Wales without a Chair of Geography, though the position in Scotland is rather less satisfactory. During the same period there has been continuous improvement and consolidation in the teaching of geography in the schools. It is a pleasant duty to salute my colleagues, the sixth-form teachers, who have so successfully implanted enthusiasm for the subject in young, vigorous, and flexible minds. To them is chiefly due the fact that reluctant Courts, Senates, Vice-chancellors, and Deans have been forced to make provision for University Geography. It has been conceded in fact that if geography is to be taught in the schools it must be learned in the Universities, and the number of students coming forward has rendered geography a subject of substantial size in the faculties of Arts, Science, and Economics.

These are well-known facts on which I need not dwell. They might appear to justify satisfaction if not complacency with the present position. Yet neither in the schools nor in the Universities can we feel wholly satisfied. We represent an idea or a group of ideas which is slowly forcing its way to recognition but which remains unpopular, or at least unappreciated, and therefore unprivileged. Ignorant or perverse individuals lurking in the backwaters of the educational stream still suffice to retard and obstruct locally. It is the burden of my argument that time is on our side and that the mental climate of our time will assist the further recognition of geography. But this is no argument for the folding of hands ; we are in no sense exempted from taking council together so that our advance may be a deliberate and considered one.

The dissatisfaction which we feel arises not from the lack of ostensible recognition but from the widespread belief among our colleagues and associates that we lack real academic status and intellectual respectability. What has been conceded is that geography has a limited use in the lower ranges. What is implicitly denied by so many is that it has any valid claim as a higher subject. Neither the Sciences, natural or social, nor the Arts receive us with unqualified cordiality. The senior geographers of this country who have made the subject what it is and to whom you and I owe so much have, generally speaking, been excluded both from the Royal Society and the British Academy. The former body admittedly concerns itself with certain aspects of research in geography through the medium of the National Committee for Geography, and it rightly makes contact through this committee and in other ways with the Royal Geographical Society. We may be grateful for the useful work it has often done, and yet find it anomalous that the University Schools of Geography are hardly represented upon it. It is right that the views of plant ecologists, soil scientists, meteorologists, and others should be sought and we would not be without their assistance. But the view represented by the constitution of the Committee is evidently the familiar one that the borderline fields of geography are interesting and valuable, but that the subject itself as a compendious whole is academically not 'on the map'.

It cannot be doubted that our case has been greatly weakened by those who have sought—and which one of us at some time has not sought—to represent geography as a subject just like other subjects with special matter entirely its own. Our confident and in the end triumphant submission must be that, on the contrary, geography is generically different from the analytical sciences and such of the Arts as pursue similar methods. This fact may not be entirely evident at the school stage, for there geography has a direct informative content largely independent of that of other subjects. The historian and the physical scientist skirt the field, and the mathematician may pause on occasion to indicate in what sense geometry justifies its root meaning. Yet it is idle to pretend that any adequate picture of the home area, the country and continent of which it is part, and the world which contains them all could spontaneously compound itself from the other subjects of the school curriculum. The crux of the question begins to emerge at the Higher School stage and to become clear and urgent in the first years at the University. The good Higher School candidate—he of the prehensile memory and the two or three supplementary answer books—is apt to think that he has 'done' Geography as his Arts fellow has 'done' Shakespeare. Then comes the 'sea-change' in which the subject seems transformed into something new and strange. The old idols and simplicities are ruthlessly shattered, and the subject seems to dislimn into a chaos of specialisms. Yet it is our case, and it is certainly our experience, that the picture finally comes whole again. The study of geography, from first to last, involves integration ; there are successive levels of geographical integration, and at each further stage more of the background moves into the picture.

The case against us primarily concerns our method. When we have made due admission of our own faults and imperfections, the faults of a young subject, unfavoured in staffing, equipment, and general regard, a subject far from fully co-ordinated as yet and lacking something of tradition and assurance, it remains clear that what is suspect is not our declared aims, but our mode of thought, and by that we must stand or fall. In rising to this challenge, I claim to speak of Geographical Science and I make the claim in deliberate protest against that narrower use of the term 'science' which is not so much current as rampant. It is no doubt natural that many subjects should seek the prestige which the name of science is held to confer. Thus political economy became economics and then economic science. 'History,' said Bury, 'is a science, no less and no more.' We need not here discuss the question at issue, while admitting that in the case of some subjects the claims made are wholly ludicrous. Thus D. C. Macintosh goes beyond the preposterous statement that theology is the queen of the sciences by inventing a system of formulae for theological concepts which can be used in a species

28

of equations to throw light on the nature of God.[1] The geographer can take up a more definite and less pretentious position. By no conceivable distortion of words or ideas can geography be forced into the rôle of an analytical science. But science also includes natural history in the sense defined by J. R. Baker,[2] as covering the investigation of objects and phenomena which can be studied as wholes, mainly by observation and, one might add, by statistical methods. Even the analytical purists have never dared to exclude the whole of this field from science. In so far therefore as the face of our planet and man in his geographical aspect are part of Nature, geography falls within the field of science, whether we are studying the natural history of land forms or of those processes and phenomena which constitute social geography.

The rôle of geographical science in education can be considered from two points of view : on the one hand, its rôle in education regarded as an end in itself ; on the other, the highly topical question of applied geography—the question of what the trained geographer can do besides teaching what he has learned.

On the first question much has been said and written, and you will expect me to do more than echo the truism that geography is a linking subject, a subject central to the curriculum and a bridge between science and the humanities. There is indeed a much more fundamental view of the matter and one which better enables us to carry the war into the enemy's camp. What are really at stake are the virtues and validities of analysis and synthesis respectively. The method of the natural sciences is analytical. The broad or crude facts of physics, chemistry, and biology are given to simple observation. Complex interrelationships and chains of cause and effect are revealed only by detailed analysis, coupled with abstraction and the exclusion of all elements deemed irrelevant. This method has proved so powerful and flexible that it has come to affect the whole of our thinking and academically takes precedence of any other. It has dominated man's thinking now for three hundred years. Yet its results in school, university, and the world at large give ground for much dissatisfaction. It is this system of thought which is the real basis of the specialisation so often deplored, by which

[1] D. C. Macintosh, *The Problem of Religious Knowledge* (Harper, 1940), pp. 202–4

[2] J. R. Baker, *The Scientific Life* (Allen & Unwin, 1942), p. 92

the student sacrifices wider horizons to travel a narrowing path designed to render him first, say, a botanist, but in the end perhaps primarily a mycologist. It is this same system which overcrowds our curricula with autonomous subjects jealously competing for shrinking time-table space. More fatally still it is the analytical method, applied pure and unrelieved, which renders our education directionless and devoid of any adequate philosophy. So grave has the position become that men of all types and outlooks have diagnosed our plight as a veritable intellectual crisis. One must agree with Arnold Nash [1] that both the Fascist and the Marxist philosophies lead to educational systems which are partial answers to the problem. They seek to give unity to education by directing it to the ends of the State or the communist ideal respectively. Neither system is without a philosophy, even if it be one which is unacceptable to most of us. Nash himself, speaking with, if not for, the Christian minority in this country, would wish to see the Christian view of the nature of man and his world re-established as the philosophical background of education. Conscious though I am that much more vigorous and radical thinking will be needed if such a point of view is to have a chance of commending itself, I range myself with this party.

On these large aspects of the general question, the geographer as such has necessarily no more, if no less, to say than any other specialist. Let us not foolishly claim for our subject that it is the happy solution to all educational difficulties. We may claim, however, that within its proper field the geographical method is an antidote to the vice of an exclusively analytical thinking. Its insistence upon regarding some aspects of Earth and Man as related wholes affords training in that integrative thinking which is so much more difficult and generally less attractive than analysis. We discovered in youth when we rashly and privily dismantled the family clock that it is easier to take things to pieces than to put them together again. Man has taken his physical and social world to pieces with exemplary zeal, but confronted with the task of reassembling them seems inclined to the view that it cannot be done or is not worth doing. It is such a reassembling of *parts* of the analysed elements which

[1] A. S. Nash, *The University and the Modern World* (Macmillan, 1943), 162ff.

geography attempts. It does not of course alone pre-empt the synthetic method ; history and the special social sciences share it too. But over an important part of the field the geographical method takes the analytical strands and gathers them to make a fabric. I cannot see that the case against this method could rest on any but two grounds, that it is impossibly difficult and (or) undesirable. We need not deny that synthesis is difficult, but the case against it must surely fall to the ground when confronted with some of the major works of geographers, few in number though they be. Mackinder's *Britain and the British Seas* (1907) stands as witness to how much more a geographer can do than geologists, historians, or others working separately or in committee. Not less do some of the volumes of the *Géographie Universelle* attest the quality of the best of geographical synthesis ; notably de Martonne's volumes on Central Europe and those on France by the same author and Demangeon published during the Second World War. It might perhaps be pleaded that these are rare and refreshing fruits from the minds of occasional mature geographers, and that the method is inadmissible because unattainable by lesser men. Yet we know how greatly the list could be extended and we perceive that the manner of thinking, if not full success in its application, is common to us all.

If then the geographical synthesis be not impossible, in what sense can it be undesirable ? Here we reach delicate and dangerous ground. We are assured by one group of thinkers that there is a true sociology of knowledge itself, that subjects arise not spontaneously or from the whim of genius but in response to human needs and social pressures. The exceptionally clean-cut separation of Sciences and Humanities in this country certainly reflects in part the mental climate of the Restoration era, when the Sciences could and did divorce themselves from politics, while the Humanities could not. I dislike some of the corollaries of this line of thinking, but it is impossible to gainsay the important measure of truth it contains. I certainly believe that the rise of geography is a response to a need, the paramount need of seeing our world and its parts as a whole. Whether the results of such scrutiny will be deemed desirable or not will evidently depend on the point of view of those to whom they are offered. I sometimes wonder whether geographical syntheses are not hindered or discouraged, not because they are ineffective, but because they reveal too much and too clearly. It may or may not be wise or possible for the analytical sciences to maintain neutrality in the world of values ; the geographer cannot so remain neutral. We do not claim of course that our craft leads to any ready or inevitable solution of disputable social and economic problems ; it may not of itself answer questions, but it at least asks them clearly. It certainly appears to me to throw a strong light on the position of geography in this country that we are so calamitously and shamefully ignorant of our Colonial Empire. Geographers have been given no adequate encouragement or opportunity to study these territories. Whether the hindrances have been deliberate or inadvertent, they have been disastrous and indefensible.

This topic brings us by an easy transition to the second aspect of geographical science in education—the case of applied geography. It is becoming very clear that there is urgent work awaiting us, both as practising geographers and as teachers. This country, in common with a large part of the world, is moving in the direction of a conscious control of its environment. Though opinions in the political sphere may legitimately differ as to the desirable mode and rate of such control, no informed mind can reject the conclusion that the state of the world or the stage of man's development calls for some measure of deliberate calculation and adjustment. The reasons for this are themselves geographical. Since Humboldt and Ritter founded modern geography more than a century ago, the population of the world has doubled itself and the whole vast range of the Earth's productive resources has been brought into use by long-continued trial and error. There is now little new land to be occupied ; the rapid westward migration of the American frontier and the more recent bold initiatives of Russian policy in settlement mark the last stages of 'filling up '. Though the population of the world shows marked inequalities of distribution, there is reason to believe that it presents us in outline with a relatively static picture in the sense that people are, economically speaking, where they belong. We know to a first approximation the world distribution of metals, fuels, and other sources of power. We have experimented widely with available crops and have bred new ones. In a word, the surface of the Earth has been developed by the mobile and ubiquitous human reagent ; its patterns have been brought out as the patterns of the photographic plate are revealed by the developing solution. As a result, spheres of occupancy, economy, and policy jostle or overlap one another, and in the developed pattern the political geographer sees pressure points and the economic geographer distressed areas. It is certain at least that there is no room for the carefree drift of *laissez-faire*. It is not the whim of tyrants or the passions of parties which have brought present looming issues to their crux, but rather tides of the spirit and surges of social process, while lying behind them both are the physical facts of man's expansion to the limits of his habitat. If this verdict seems to ignore the intimations of an approaching atomic age, it can hardly be doubted that any such new major factor will complicate rather than simplify earth patterns and human problems.

It is against a background no less widely viewed than this that the geographer must find and follow his rôle. It concerns three distinct levels of action. In the first place, it has plainly become necessary to maintain the supply of trained specialist geographers. It was formerly a grave weakness of our subject that it led to no career but teaching. We are most of us teachers here and not ashamed of our calling, but we are able to acknowledge the small element of truth in the cynical dictum 'those who can do, those who can't teach'. Perhaps no subject requires further vindication than that it contributes to general enlightenment, but it is a healthy and stimulating feature if its crafts and skills are directly applied in the field of practical problems. Such a position has been fully attained by geographers of late. There are many aspects of a planned economy which lie beyond their interest and competence, but at least, in a small country like our own, land use is basic to the whole planning structure. There *par excellence* the geographer comes into his own. I shall not attempt here to assess the value of the geographer's direct contribution to the war effort and the not less urgent peace effort. It is sufficient to remark that at the moment the demand for geographers exceeds the supply and university staffs have been seriously depleted by the needs of government service.

Even more important than the training of specialist geographers is the diffusion of the distinctive geographical method among administrators. My own experience in this field is limited and various friends and colleagues know it much better. But I have

recently seen enough to realise how vital to our times is a knowledge of modern geography among the able and disinterested men
who occupy senior posts in our Civil Service. They are willing
to accept geographers as specialist colleagues, but show no
awareness either of the vast available body of geographical data
or of how it can be used. To some extent the obstacle is the
inimitable intellectual snobbery of a certain type of 'Arts' man.
A. L. Rowse has recently averred that 'one may be a cultivated
man without knowing mathematics, chemistry, or engineering
for those are specialisms. We expect the technicians in question
to know them and to do our sums and sanitation for us.' [1] By
contrast, to him a sense of history is part of the 'self-awareness
of our environment'. To this we may agree, with the important
qualification that part of the environment is geographical and any
awareness which ignores this is to a large extent unawareness.
Many of the products of our dominantly literary culture are
woefully ignorant of the simple physical facts of environment,
how soils form, how plants grow, and where underground water
comes from. What is required to repair these grave defects in
mental equipment is not that largely imaginary subject 'general
science' but geographical science in the widest and most en-
lightened sense of the term. For it is such simple facts, processes,
and phenomena as related to area—to administrative boundaries
and competing interests and uses—which enter continually into
practical governmental problems. 'Geo-politics' is a word so
debased by usage as to be repellent and perhaps misleading.
Yet great fields in both the internal and external spheres of
government are concerned with a species of applied geo-politics.
Until this fact is recognised and acted upon the difficult and
delicate work of planning will be gravely hindered. There are
not enough geographical specialists to carry unaided the burden
of the necessary co-ordinated thinking, and much of their effort
is spent upon educating their administrative colleagues, many
of whom reveal gulfs of ignorance for which an intermediate
student would be ploughed without hope of reprieve.

If awareness of environment on the part of the governors is
a prerequisite of any sound government in the world of today,
how necessary it is also for the governed, at least in a democracy.

[1] A. L. Rowse, *The Use of History* (Eng. U.P., 1946), p. 191

This I believe is the aspect of the matter which concerns the greatest number of teachers. I ventured, in speaking at our Conference last year, to commend the study of such documents as the 'Greater London Plan' because they mark the initiation of great experiments which will take generations to complete. A year has passed and the Plan is in course of application—or, as some may feel, of misapplication. Yet there has been in my judgment all too little informed public discussion of its implications. The children in our schools must be taught to comprehend and to appraise the world which these social documents are seeking to evoke. Theirs is the heritage and with them will lie the ultimate decisions. Otherwise they will find themselves, to vary a now famous phrase, 'more planned against than planning'. Is it not a plain duty to clarify, strengthen, and if necessary remodel our geographical teaching to fit such needs and occasions?

The chief question before us here is the part to be played by geography in the Secondary Modern Schools ; of the Primary and Grammar Schools I will say nothing, for though no doubt improvements are possible, excellent geographical work is being done in both. For the Secondary Modern Schools we have no agreed curriculum, no decision on the place of geography therein, and no clear idea of its content for these purposes. All that seems to have been done is to enunciate some high-sounding and unexceptionable principles about merging subjects and avoiding an academic outlook. The geographer, by temperament and outlook catholic and no believer in subject autonomy, will not be disinclined to co-operate with historians or to join the family of Social Studies. There is however a real risk that, regarded by others as a junior partner in such mergers, he may find his subject merged out of existence. Most of the attempts I have seen to combine history and geography give exceeding over-emphasis to history, tactfully tincturing it with a few indifferent maps. Such will not serve, nor is there any profit in pretending that the rudiments of sociology are a substitute for geography. What I chiefly fear, if the narrower humanists gain control, is that the physical elements of geography will be unduly subordinated. I will venture the opinion that there is a pressing need for a strong joint committee to consider this problem, a committee on which the Royal Geographical Society, the Geographical

Association, the Institute of British Geographers, and Section E of the British Association should all be represented.

May I offer finally a few concluding reflections on the argument I have presented ? It is not of course my own argument ; it has been stated and better stated many times. But the time has not yet come when it can be taken for granted ; it needs frequent restatement with all the emphasis we can command. For the world is not yet wholly convinced ; it often does not recognise modern geography when it sees it. Some of the reviews of that recent admirable publication of the West Midland Group —*English County*—were inclined to take the line, 'How much better this is than mere geography'. It is indeed better than mere geography, whatever that may be ; but it is largely the work of geographers who have obviously outgrown the limitations of 'mereness'. The cordiality with which such work is received enforces a further point. It is necessary to restate the geographer's case, but it is more necessary still to demonstrate it. A great responsibility rests upon those geographers who have the opportunity of demonstrating their craft in action in the field of planning. Not less is it incumbent upon us to give earnest of the value of geographical syntheses by making independent studies ; we are judged not by projects but by performances.

I have referred to the need for such studies in the Colonial Empire, but their failure to appear reflects lack of facilities. Less easily can we escape blame for the fact that the regional geography of Britain is still unwritten. It is nearly twenty years since the various university schools produced, under the editorship of Professor Ogilvie, that useful but necessarily imperfect volume *Great Britain : Essays in Regional Geography*. As every contributor would agree, there is scope for much more than this. We have in fact been guilty, amongst ourselves, of a lack of faith in our own regional method, deeming it worthily applied by immature students in theses which rarely see the light of day. I am not forgetting the *Land of Britain* series which partly, but only partly, fills the gap. For myself, I look to a revival of H. R. Mill's project, illustrated by his paper on 'A fragment of the geography of England' published in the *Geographical Journal* in 1900.[1] This has served as the prototype for diploma and graduate theses, but

---

[1] *Geog. Journ.* 15 (1900), 205–27, 353–78

it has never been followed up by senior geographers equipped with full knowledge of their subject. I would definitely include among the suggestions I have ventured to make this evening, that Mill's proposal for an adequate geographical description of this country be taken up again as a research project, as a means of vindicating the geographical method and forwarding the ends of geographical education. I have said that the world is not wholly convinced by our case : when I think of the fate of Mill's scheme, I am impelled to the conclusion that the Royal Geographical Society itself was not wholly won over by the arguments of the academic pioneers. It is now sixty years since Sir Halford Mackinder—a young revolutionary, as he then described himself—read his paper here on 'The scope and methods of Geography'.[1] It was from this platform in 1931 that he recorded [2] 'the *obbligato* of seafaring language which then came *sotto voce* from a worthy admiral, a member of the Council of the Royal Geographical Society, who sat in the front row'. If I have heard no such *obbligato* tonight, I am far from attributing it to my superior powers of advocacy ; rather may I say that the years between have created a different mental climate which may perhaps give us the opportunity of reaping where our predecessors sowed. I remarked at the outset of this address that time is on the side of our subject—or, as I would prefer to say, of our point of view. But this can only be so if we rise to the measure of our opportunity. We have a thing to say ; what is now required of us is not to declaim it in theory but prove it in action, in the classroom, in the laboratory, and in the wider world which is both the object of our study and the appointed field for its application.

[1] *Proc. R.G.S.* **9** (1887), 141–74
[2] *Report Brit. Assoc. Adv. Sci.* (London, 1931. Pres. Address to Sect. E), p. 97

3

On Taking the 'Ge-' out of Geography

The Importance of the Study of 'Physical Basis' at all Stages of Teaching

MY TITLE might perhaps have proved a little enigmatical had I not offered a rather clumsy translation of it in the sub-title. From that you will rightly gather that I propose to speak about the rôle or status of physical geography in our work. This may possibly seem to you an old and threadbare topic. 'Who,' you may be disposed to ask, 'has ever questioned the importance and relevance of physical geography, in due measure, at all stages of instruction?' I shall answer this question as my argument develops. For the moment I wish merely to add that neither title nor sub-title expresses as exactly as I could wish the group of questions which I have in mind. Central among these questions are those raised by current discussions on integration within the school syllabus. There has, as you know, been much talk of the effective merging of history and geography and of the emergence of a new subject, or at least a new approach, under the heading of Social Studies.

Geographers must inevitably be critical of some of the tendencies of this recent line of thought. We must certainly avoid at all costs that characteristic indignation of the ' vested interest' which was incurred by our own subject in its earlier days. It is no fit argument against any new approach that you and I have sunk intellectual capital in our own fashions of thought and teaching. That way lies stupid and unworthy reaction of a type regrettably familiar in the educational world. None the less we are entitled to our reflections on the promised development of what, for simplicity's sake, I will call a new subject which overlaps with, if it does not entirely pre-empt, much of the field of geography.

One may arrive at two alternative conclusions from a study of the hopeful and enthusiastic books and syllabuses of the supporters of social studies. On the one hand, it may seem that they have rather naïvely rediscovered the geographer's idea or ideal and published it under a different name. But while it is impossible to resist entirely the impression that the reformers are engaged in fact in the old British sport of 'teaching their grandmother to suck eggs', this is far from the whole truth. There is, potentially, more in social studies as currently conceived than geography can ever properly have attempted to cover. If it has in fact attempted to cover it, it is because the gap has existed and no-one, the historian included, has made much shape at filling it.

The alternative view to which we may come is that the new synthesis is designed to embrace geography, with the *ge-* largely omitted, together with appropriate borrowings from history, elementary sociology, and economics. We can see clearly how such a synthesis arises. The study of societies is not exclusively the business of the geographer. The historian will not consent to his exclusion and the economist and others will seek entrance, at least in the higher stages of the work. It could all too easily result that the field of geography in the view of these colleagues might contract to that of physical geography alone. They might be willing to leave us the physical environment if all or most of the human environment passed to their charge. Perhaps I may say that, as a geographer reputedly interested in the physical side of the subject, I should be in no sense satisfied with this ; nor, I am confident, would most of you.

The villain of the piece in much of the unclarified thinking on this matter is the over-worked word 'environment'. It can signify so obviously almost everything as to come to mean almost nothing. H. McNicol [1] in a valuable and stimulating book, distinguishes the physical, human, æsthetic, spiritual, and practical aspects of environment, and, though the categories are not entirely mutually exclusive from a purist standpoint, they will serve. The physical environment he and other writers naturally delegate to the sciences and, as he puts it, 'certain aspects of physical geography'. The human environment, involving the relation of man to man, also involves geography in other aspects, together with history as the illuminator of 'heritage',

[1] H. McNicol, *History, Heritage, and Environment* (Faber, 1946)

39

and literature. Thus the field of geography is disrupted. It is fair to add that the solution of the syllabus problem propounded by McNicol does not in fact go so far as this. Following lines laid down by Dr F. C. Happold in his well-known works, he regards the five-fold study of environment as affording the basis of an *ideal* solution, but he decides, no doubt for practical reasons, in favour of a 'second best', which unites elements of history and the social sciences, but leaves the geographical elements out. Any satisfaction which we might feel at this courteous refusal to reshape our field for us is tempered by such doubts as I have already expressed as to what it is, in fact, which is left to us. I will not labour the point. The tradition in which we have all been reared in this country is the unitary view of geography which gives due place to both physical and social aspects. If the latter aspects were lost to us we should finish in a position far from that for which the great founders of our subject strove.

All this you may find a rather curious introduction to my promised theme of the importance of physical geography. To make my argument clear, therefore, may I ask what I think is the necessary important question. How comes it that historians, economists, etc. think that they can thus pre-empt or take over a large part of what we have called human geography? They propose in effect a new synthesis which takes part, but not all of what we have to offer ; they reject our own synthesis. The fault for this, I believe, is our own. We have not preserved the balance between physical and social ; we have rendered the former of so little account that it seems that it can be brushed aside without serious loss, or relegated hopefully to physics, chemistry, and biology. Our fellow humanists have found us in fact so preoccupied with matters that they regard as properly their own that they feel they can subsume geography, as we have made it, under some new and different heading, either dispensing with our services or at best including us as a junior partner in the enterprise. Briefly, if tacitly, their criticism I believe is this ; if geography has in fact become nothing more than social studies, why not say so and at the same time admit to that field those whose right to teach it is at least as great as, if not greater than, that of the geographer himself.

It is needless to say that I do not accept this view, but I have a certain sympathy with it when I see my subject increasingly diverted to the study of Man rather than Land, using that term in its widest sense. Geography has always been subject to the danger of pitching its claims too high and its realm too wide. It cannot and must not claim the whole of 'social man' as its province. Geography, as Vidal de la Blache says, is not the study of man, it is the 'study of places'. Or again as Sauer and Leighly put it 'Geography has never been the science of man, it is the science of land, of the earth's surface.' These may be unwelcome statements to some of us, but if they are, it is because our thinking has become confused and our real aims obscure. They are at least authoritative statements, and prove that it is not some curious personal heresy which I am trying to urge upon you. Do not for a moment suppose that they imply that man is to be left out. It is man who gives significance to places and it is man who is the chief developer of the earth's surface, developing it as the reagent develops the photographic plate. Many of us probably give a tepid assent to these doctrines without perceiving their full implication. And part of that implication is that there can be no geography without physical geography.

So we come back to that unexceptionable sentiment for which any amount of lip-service can be won that a due place exists for physical geography. But it all depends on 'due'; in fact, physical geography is shamefully and consistently neglected and some of the reasons for this are not far to seek

In the first place, it has for many been deliberate policy controlled by the dogma that physical geography should be introduced only incidentally or as needed in the course of regional work. This was certainly the view of my own training-college days. We were led to believe that the 'capes and bays' geography had given place in its turn to a phase of undue emphasis on the scientific or physiographic aspects, and that these, at their worst, were little better than 'scientific capes and bays'. The essence of the method then recommended was that though it might be logical it was not psychological to deal formally with physical background before turning to human response ; in short it was bad teaching. This doctrine had some very real theoretical and practical merits ; it is, for example, a sound specific against determinism, but I have latterly reached the conclusion that it has done harm as well as good. It is indissolubly linked in the history of the teaching of the subject with the breathless rush to

41

'cover the world' in our courses. This meant that there was little or no time, anyway, for the formal teaching of physical geography, and we were glad enough to find necessity made into a virtue by the prevailing creed. We brought in the physical geography where we could not do without it, or used it, as recommended by our mentors, to beguile the tedium of hot last periods or Friday afternoons. It was and still is possible to make an outline survey of the world with only the slightest reference to real physical geography. More concerned with consequences than causes our categories are the simple ones of land and sea, mountains, plateaux, and plains, together with the broad facts of climate and vegetation. Some of this you may call physical geography if you like, but on the one hand it is so simple as to contain little of interest in itself and on the other so generalised as to be rather repellent to young minds. The vivifying elements of significance behind it, derived from geology and meteorology are ruthlessly excluded and, in view of all the difficulties, perhaps inevitably so. It is instructive to turn the pages of any one of the excellent General Regional Geographies of the world at present current. One is lucky to find as much as a tenth of the text devoted to basic physical material, even including so-called mathematical geography. There are the usual brief and rather arid summaries of the absolutely essential trifles of geology and meteorology ; at this level physical geography is peculiarly dead, and one is not surprised that it is reckoned of little account. In the succeeding regional accounts, in harmony with the dogma, physical matters receive scant and purely *ad hoc* mention. On the geological side there may be a section across the Weald or the Pennines and a reference to the Niagara Falls, the Fall Line, the Rhine rift-valley, or the loess of North China. As to the real mechanisms of the atmosphere the text is generally quite silent.

I would be the first to admit that there are very real limitations to what one can or should teach under these heads on this scale of treatment. But since this 'world geography' has come to take up so much of our time and thoughts, the subordination of physical interests here sets the fashion throughout our work as a whole, and plays directly into the hands of that considerable minority of teachers who find physical geography difficult and distasteful. To such the social aspects of geography seem easy and pleasant. So they are, if we skate circumspectly on the surface of things as we do with our 'man in many lands' in the case of young children. But in fact, of course, the social side is much more difficult of real comprehension, and I have had ample evidence over many years that the real degree of understanding achieved, even at Higher School level, is gravely limited. 'On a Norfolk farm', 'Down a Yorkshire coal-mine' or 'Around the Ford works at Dagenham'—these are familiar themes to us and well enough, in their way, providing we remember that the real intellectual problems even at the upper school levels concern not the geography but the economics of such cases. One does not get any real insight into how the economic system works from such instances as these, treated descriptively. If geography is to be anything more than an easy introduction to everything else, we cannot substitute this type of work for real study and thinking, beyond the veriest introductory stages. It is certain that no case can lie against the teaching of elementary physical geography because it is difficult as compared with real social and economic geography.

An alternative plea is that children do not like physical geography. This I know to be demonstrably untrue. I have taken considerable pains to amass evidence on the point, and before I listen patiently to the quite frequent objection I require to know what the objector means by physical geography and how he teaches it. I concede that before the age of generalisaton is attained, few things are more sterile than the generalisations of climatology. Ocean currents, too, are really statistical abstractions. All our work on the envelopes—air and ocean—demands considerable powers of conceptual thinking, and it is fatal to introduce it too early. Not so, however, with hills and valleys, rocks and streams, soils and plants, which lie at our doors or not far from them. The regrettable thing is that schematic presentations of air and ocean circulation are easy to the teacher, because they are neatly contained within the book. The accessible realistic physical geography must be sought not in the books but shod with a pair of stout boots.

A last attitude which is sometimes displayed towards physical geography is that much of it is irrelevant to our theme of 'The Earth and Man'. It was tersely expressed by our old friend in this University, Professor L. L. Lyde, who was wont to say that much of physical geography was 'mere morbid futility'. It is fair to own that few either of the staff or students in the great school he founded acted upon this dictum, which was perhaps not intended to be taken at its face value. The fallacy in the comfortable doctrine that physical geography is irrelevant is a natural outcome of the dogma we have noted. Naturally enough, if you borrow from physical geography only when it happens to suit your argument its rôle will appear spasmodic and incidental. As a possible example may I note that twice and only twice in the illustrative maps of that stimulating book *Geography and World Power*[1] does it appear that the land surfaces of the globe have a varying geological constitution as well as mere form and extent. There is a map showing the sandstones and limestone areas of the Nile Valley, and one indicative of that hardy perennial, the loess area of North China. To the uninitiated reader it might appear a little strange that the actual 'stuff' of the earth's face only prominently entered the game in these two curiously different and widely separable cases. In fact, of course, Mr Fairgrieve is right within the limits of his argument. At this scale of treatment rock character is rarely a major determinant of human action or reaction. If I had attempted to write such a book, a task for which I have few qualifications, I might well have found, by virtue of my interests, some other cases of the significance of rock pattern and accorded it a bigger rôle in 'world power'. I might have spoken of the Deccan Traps and the black cotton soils, of the vast emptiness of the Pre-Cambrian shield of Canada, and of the curious antithetic influence of auriferous and alluvial rocks in the Spanish colonisation of the New World. Or I might have traced the great Hercynian zone in America, Europe, and Asia and reflected on its recent past and its probable future. But that is not my point. Mr Fairgrieve makes his own legitimate selection of world-shaping factors ; he would not, I know, deny the relevance of others. In any case it is only by studying physical geography that we can determine how and where it is chiefly relevant.

So far my approach has been critical. I have dared to challenge the time-honoured dogma that physical geography should only be taught incidentally. I have rejected the alter-

[1] by J. Fairgrieve (U.L.P., 1927)

native pleas that it is difficult, or uninteresting to children, and I have declined to admit that it is irrelevant. Let us now pass from the critical to the constructive and state a positive case for the teaching of physical geography. But it is not narrowly 'physical' geography that I want to commend, but rather what might be called natural history geography—or, more shortly, natural geography. This extension is necessary because we are in danger of forgetting that there are three levels of integration in geography—physical, biological, and social. Too often we elide the middle term in our eagerness to develop the social aspect. 'Natural geography', if I may so term it, has its independent place and distinctive rôle in the curriculum because it is the starting point of other subjects and because of its own cultural value. We are right to reflect that the children in our classes are the citizens of the future and to make our contribution to that end. But their lives will contain more than civic and political aspects, however important the place of these. Our grammar-school classes also contain the potential field scientists of the future, and here and elsewhere a host of others who at least will travel in their own country and spend holidays in it. By what right can we withhold a knowledge of how our terrestrial home is constructed, the meaning of its scenery and of the patterns of its sky. The world contains not only factories, farms, railway sidings, market-places, etc., but the blue hills on the skyline, the winding valleys which traverse them, russet bracken-covered slopes, and heather fells and the ever-changing incident of long and varied coastlines. In these aspects our terrestrial home is a thing of beauty and of interest. It is a monstrous perversion of natural geography which sees in it nothing more than an occasional prop to social geography—a perversion arising from what I would call the 'humanist fallacy'. As children we perhaps were familar with the words of Isaac Watts :

> I sing the Almighty power of God,
> That made the mountains rise,
> That spread the flowing sea abroad,
> And built the lofty skies.

Even if we reject, and not all of us would do so, the theological implications of the statement, we may yet be disposed to admit that at least we lose some sense of the majesty and coherence of nature if mountains are only barriers, and seas the occasion for

45

trade routes, and the 'lofty skies' the mere source of monsoon or other rains. Man's terrestrial home is beautiful and interesting in its intricacies and harmonies beyond his power to make or mar it. It was not wholly gain when that fine old subject 'physiography' passed from the educational scheme as 'economic man'— mutation of *Homo*, miscalled *sapiens*—rose to dominance in the earth. It contained not a few scientific 'capes and bays' better forgotten, but it did much to inculcate a sense of the unity of nature.

I would therefore urge that we have in no sense filled and finished our rôle as geographers by careful studies of how bootlaces are made, or where Ovaltine comes from, and so on. Such topics are well enough in their way, but other students of society will have words to say here and they may on occasion prefer their words to our own. No-one, at the school level of instruction, will seek to dispute with us the field of natural geography. Do not seek to assign it to the laboratory scientists, they have their own approach, which is not ours. It may on occasion happen that you are fortunate enough to have a biological colleague who is not only a field man but an ecologist rather than a systematist. If so, you have a brother of the craft ; he is geographer under his skin so far as concerns his basic way of thought, and you can commit to him what otherwise you might perforce have to attempt yourself. But you will certainly have to teach the elements of geology and meteorology yourself, counting it a privilege not a hardship to do so. There has been widespread comment in recent years upon the disastrous ignorance of what may be called the geographical aspects of geology among the educated classes. I have never been in favour of teaching geology as such in school. It is properly introduced by the geographer, both because he needs it for his own more special purposes and because by his outlook and interests he is generally the best-equipped member of the team to show its significance in the intellectual and economic life of mankind. It is a permissible reflection that the vast majority of the geologists passing through the British Universities obtained their first introduction to the subject from teachers of geography. If geologists were more mindful of this they might, I think, sometimes treat the sister subject with more courtesy and charity.

You may object here and remind me that I have already condemned the notion that geography should serve as 'an easy

46

introduction to everything else' on the social side. My reply is that the two cases are in no sense completely comparable. However slight your geological pretensions you are probably the only person on your staff with any knowledge of geology, and the only one therefore who can show how such knowledge enlivens, say, a railway journey, gives significance to hill and valley, and illuminates 'the long results of time'. For myself I should wish you to regard all these aspects as central to your proper work ; but even if you do not, it remains true that they have their own interest and value and you alone can introduce them. As we have seen, this is by no means the case over wide sections of the social field. In them, too, you may be interested, and concerning them you have a word to say. You will find it worthy of reflection that this word itself will often derive from your knowledge of natural geography.

I do not doubt that many of you agree, in part at least, with the line of thought I am pursuing, and like myself you may pause sometimes and wonder how it has come about that elements so sane and wholesome in themselves and so manifestly proper to a full education have tended to fall into the background. My own conclusion is that, in this country, much is attributable to the over growth of urban life. Among the many controls of environment is that exercised upon subjects and their teachers. Geography has become, I think, not so much 'man-centred' as 'town-centred'. Escaping from the rigours and fallacies of the determinist creed that man is the creature of his physical environment it has fallen into another error, akin to the famous and flourishing doctrine of dialectical materialism, which regards the economic as taking precedence of everything else. I dare not say that geography has become Marxist ; to do so might win me extravagant adulation in quarters from which I should not welcome it ; it would as surely bring violent reprobation, and deservedly so. But there is an element of truth in the idea ; it is a particularly nasty and narrow view which regards the lands and landscapes of our earth as simply and solely the theatres of economic processes. Here I join hands with those geographers, including I am sure our last President, who claim that you will never understand man if you cut off from his past everything that happened before the Industrial Revolution. But it fits better my own beliefs and my present thesis

47

to say that we shall never understand Land in relation to Man if we view it too exclusively from the standpoint of the Old Kent Road, Tyne-side, Brightside, or the Rhondda valley. It is admittedly an important standpoint, but one to which we in Britain are apt to give an undue prominence. Time may well correct that ; already as we know there is other writing on the wall. Meanwhile, let us beware of over-emphasis at the expense of natural geography.

In so doing we can rid our minds of any fear that we shall be leaving man out. Here I reach what for me is the secret of the whole matter : that as you study natural geography man comes inevitably into the picture and in his right place as part of the terrestrial unity. The essential correctness of this view is shown when we recall that it was in just this way that geography originated. So it was certainly with Humboldt and Ritter, and so it was with our own Mackinder, chief pioneer and prime fashioner of British geography as we have come to know it. Great humanist though he became, his first degree at Oxford was in Natural Science, and when he made his famous pronouncement on the 'New Geography' at the Royal Geographical Society in 1887, he based his most telling example upon the influence exerted on its geography by the geology of southeast England. In the same way Geddes was a biologist and Grenville Cole, who fully grasped the geographic thesis, was a geologist. Geography has, it is true, been approached from other sides, notably by historians, and its debt is not less to these. But I am sure that it can be fairly maintained that in all three of the 'centres of action' which gave birth to British geography, London, Oxford, and Edinburgh, natural geography, the part, generated geography, the whole. The old road will not mislead us ; it is the one which most readily gives unity and coherence to our subject.

I think that, by now, whether or not you agree with the views I am expressing, you will desire to recall me from theory to practice. To that I most willingly turn.

Our subject is necessarily enclosed within the four concentric circles of neighbourhood, region, home country, and world. It will be clear from what I have said that I regard the essential division here as between the first three and the last. I have implied that only in the accessible realms of the actual ground, of large-scale maps supplemented as far as may be by fieldwork, can natural geography really be taught. I claim that it should so be taught and that therefore the local, regarded as including neighbourhood, region, and home country, should figure in every year of the course. Preferably it should occupy about half the time and certainly not less than a third. Is it so unreasonable or impossible that say two periods in three weeks should be given to these aspects ?

And what of Social Studies ? These too will find some of their chief opportunities in the local field, but if we devote all our necessarily limited 'local' time to social aspects then natural geography will fade from the picture. If the subject Social Studies is to come into the curriculum, then in the early stages I would agree that it does not matter whether an historian or geographer teaches it, but it is clear that neither one nor the other will be content if the subject is to be but a condensed substitute for the subject it supposedly replaces. We may leave it to the historians to make their own case, but the geographer must certainly be given further time to deal with the *ge-* in geography. At the higher level, social studies must certainly, in my view, figure, if at all, as an additional subject, and is then best taught by an economist. Academic economists are not in general favourable to the teaching of economics in school. What is really required is what Professor J. R. Hicks in his admirable introductory book, *The Social Framework*, calls 'social accounting'. A glance at any such book will quickly show the geographer how far this field diverges from his own. Even if he made shift to teach these aspects he could not but be aware that he was perforce ignoring the greater part of his own synthesis. Yet only, I believe, in such terms can Social Studies at the fifth and sixth form levels be given any reasonable disciplinary fibre. Without economics, the only fully developed member of the social sciences, social studies form a circle which, to borrow a phrase from Winston Churchill, would be 'fluid, friendly, but unfocused'. The required focus is not our focus and without despite to the merits of the new view we must maintain our own view of man and the earth.

I conclude, therefore, that nothing would be lost, though little perhaps gained, if in the elementary stages both historian and geographer gave part of their attention to what might be termed Social Studies. At a higher stage there is much to be said for Social Studies in the place of Civics or Economics. But there it cannot even partly replace Geography, properly so-called, though the geographer could make some telling contributions to it. But let us not, drunk with the wine of new doctrines, claim it as either chiefly or entirely our own, or accept it as a substitute for our own synthesis. If we do, then the *ge-* will certainly be taken out of geography, and the baby thrown away with the bath water.

Reflections on Regional Geography in Teaching and Research

I

BY REASON of its nature and the circumstances of its rise and growth, what may be termed the 'apologetics' of geography necessarily embraces two distinct aspects. Ever and again there comes the need to explain our aims and methods to neighbour specialists and to laymen. More rarely, as on such an occasion as this, we may take counsel together on the internal problems of our subject. It is rarely indeed, in this country, that one faces an audience of which the members realise clearly both what geography is and what it might be. Here at least much may be taken for granted, and there is no need of eloquent expostulations on the value and validity of our point of view. Despite all differences of interest and emphasis amongst us, we are the inheritors of a consistent and continuing tradition such as is necessary for any healthy academic discipline.

I have long since lost faith and interest in attempts narrowly to circumscribe and strictly to define the field of geography. The muttered sneer that 'so-and-so is not really a geographer' is a red rag to me ; in effect, if not in intent, it is generally a pusillanimous, muddle-headed, but tragically effective method of fouling one's own nest. But to affirm this is not to deny that our subject has centrifugal tendencies and that convergence from our wide periphery towards our central objective is strenuously to be sought. And I join with many better men of former days, among our founders and pioneers, in claiming that this central objective is regional geography.

I did not always think thus. The original organisation of the Joint School at King's College and the London School of Economics made an exceptionally clean separation between systematic and regional geography, in that most of the former

51

was allocated to King's College and all of the latter to its consort. Cartography, physical geography, and bio-geography, together with the history of geographical discovery and much of historical geography fell to my colleagues and myself, while in the regional teaching, together with economic geography, we did not share. I came to the subject as a meteorologist and taught this branch in isolation from its regional applications, of which I was then more than a little contemptuous. To many of us it seemed that the student treatment of regional geography was little more than a regurgitation of factual gruel, having no claims to the status of either science or scholarship. Fuel was added to the flames of resentment by the calm assumption of regional colleagues that theirs alone was the true geography, and that a good performance by students in regional papers was a special sign of intellectual distinction outweighing a sometimes palpable ignorance of systematic disciplines. I may perhaps recall my temerity at the London meeting of Section E of the British Association when, in reading a physiographic paper [1], I ventured to join issue with the great Mackinder himself, President of the Section and founder of the school in which I taught. His view of the subject, I said, seemed to reduce the rôle of the physical geographer to that of a humble, if necessary, laboratory assistant—a mere server at the regional altar. I did him less than justice ; his reply was magnanimous, disarming, and wholly characteristic of his great width of view.

Though my own views have now changed, those early years in the joint school at least enable me to understand, in part, the gravamen of the charge against regional geography as commonly taught ; nor, even now, do I find it by any means wholly baseless. That there is an issue to be resolved is apparent from the very varying practice amongst us in the weight given to regional teaching. I think it can never have bulked more largely than in the former regulations for Honours in Geography in the University of London ; these prescribed five specific regional papers out of a total of nine. These regulations have recently been substantially modified, but it is too early as yet to judge of the success of the reform. What I have to say today is based on experience of the former system in London. It would clearly be a grave abuse

[1] Included as Essay No. 9 in this book

of my position to criticise the arrangements of others. If anything that I say finds an echo in the experience of your own departments, I leave it to you to note that the cap fits ; if not, you will at least have the perennial satisfaction of thanking your gods that you are not as other men are.

2

It is needless to spend time here in stating and defending at length and in general terms the aim of regional geography. It is to gather up the disparate strands of the systematic studies, the geographical aspects of other disciplines, into a coherent and focused unity, to see nature and nurture, physique and personality as closely related and inter-dependent elements in specific regions. It involves no nice distinctions as to what a region really is. We had better, I think, in any case, abandon pseudo-organic analogies, and we must recognise that there are many generic types of region, but for regional geography in its present stage, specific regions will suffice.

To that great majority of us who have been brought up to a belief in the essential unity of our subject, the aim of regional geography, so expressed, seems not only defensible but of great intellectual attraction. Yet we do well to note the lines of possible attack upon it. In the first place comes the charge that we are seeking a unity which does not exist. I referred just now to the 'disparate' strands of the systematic studies ; if they were strictly disparate in the sense of 'incommensurable', then, evidently, our hope would be vain. In some moods it may seem indeed that the complex interrelated unity for which we seek is an aspiration of faith rather than a fact of observation, and this is certainly a view to which our critics, at least in this country, are notably prone. From the early days of the Royal Society the separation between the sciences and the humanities in Britain has been radical. We are apt to claim it as a virtue that our own subject ignores this time-honoured cleavage, but the penalty we pay for our point of view is effective exclusion from both circles of the learned. Yet I am convinced that we should surrender our attempt at 'binocular vision' only at our peril, selling our birthright for a mess of 'one-eyed' specialist pottage. None the less we do well to admit that we have taken on an exceedingly difficult task and

5

to be modest in the claim we make for the progress so far achieved.

This leads me naturally to the second and more serious criticism which comes both from outside and inside the subject. If it be admitted that a synoptic view of very diverse inter-acting factors in a region is conceivable, if it be allowed further that it is eminently desirable, there still comes the question, ' Are you, the self-styled geographer, at once geologist, meteorologist, biologist, historian, economist, and sociologist enough to command and to achieve the synthesis ? ' Implicit in this is the charge so often made against us of superficiality, and it is a vein of the same style of thinking which so often declares, with such complacent finality, that geography is a post-graduate subject. The reply to this last version is evident enough. If Cinderella is to stand last in the queue of her beautiful sisters she will, indeed, lack for suitors. I can see no reason why any adequate number of graduates in other subjects should turn to geography either through choice or circumstance. By the time that they have suffered the rigours of one-track specialisation, they will, from our standpoint, have donned blinkers and adopted the bias that will generally last them to the end of their days. Thinking synthetically is no mere 'food for babes' or something easily or almost contemptuously added, as an afterthought, to a customary training in analytical technique. In the words of Fairfax to Jack Point, "Tis an art in itself and must be studied gravely and conscientiously'. But this granted, let us own that it is quite natural that narrow analytical specialists should look askance at a scholar attempting to do more than one thing at once. And before we rebut the stricture let us note that we ourselves tacitly assent to the charge whenever we give way to untimely specialism. Long years of active debate and hard experience in our Joint School have convinced me that an historical or physical geographer is not, and cannot be, a 'thing in himself'. Until he has faced, and, in some degree, mastered the difficulties of regional study he is often hardly geographer at all.

The most serious criticism of all comes, or should come, from ourselves. It appears to me very difficult to deny that, in regional geography, we have not as yet often risen to the height of our own argument. I feel gravely dissatisfied with much that passes for teaching in regional geography—my own not less than that of others. We all too easily become content with a less than second-best. We are aiming, we say, at integration, but there are evidently different and successive levels of integration. At the school stage, fifty years of sincere and devoted effort, following the initiative of Mackinder and his associates, have produced a recognisable and definitive subject. At this level one must in some sense 'cover the world', though the breathless effort to do so within the allotted confines of time is perhaps the greatest single limitation upon the depth and discipline of school geography. Certain it is that the regional synthesis must be in simple terms comprising the barer facts of relief and climate as background to a survey of human economies. So far as it goes the synthesis is successful, exercising a real attraction on younger minds and inspiring them with the desire to go further. And the ex-service students of recent years, who, if they formally know less, have at least seen more than their younger associates, echo the cry of Kipling's time-expired veteran :

> For to admire and for to see,
> For to be'old the world so wide,
> It never done no good to me,
> But I can't drop it if I tried.

How, then, are we to raise the level of integration above that of the school stage ? It is of no avail, with the rather fuller time allotted, merely to convey more information, a fuller factual digest, than school conditions permit. The vice of the geographical method is that generalisation so often precedes, or even replaces, careful analysis, seeking simplicity by ignoring actual complexity. Let us take as an example the North Italian Plain. The school texts sketch, with only minor variation in content and emphasis, an outline of its towns, routes, and occupations. If we turn to Dr Marion Newbigin's more advanced treatment we find that it comprises some 5000 words, equivalent to a short paper or an hour's discourse. This is, indeed, an excellent account by one of the best-equipped and most successful writers on regional geography of our generation. Yet the account does not greatly transcend the school level. It can be read, nay worse, it can virtually be learned by heart, by Higher School candidates who have never seen a single large-scale map of the area. Suppose, then, that we go a stage further and construct a block of the Austrian Staff maps of the region. Immediately a host of interesting patterns appear and an equal number of problems present themselves, each leading seductively into some specialist preserve. The prograding Adriatic coast would, no doubt, touch a chord or two in the breasts of some members of the Cambridge School. I myself am curious concerning the limits of the Pliocene trespass in the foothill zones, this not as a geological, but a geographical, fact, and one likely to be significantly legible in the landscape. In other moods we might speculate on the vicissitudes and fate of Roman Aquilaea, on Venice, with its gorgeous East formerly held in fee, on the urban geography of Milan, the significance of hydro-electric power, or the implications of autarky in agriculture. As we dig into the regional problems we should, in any case, become very dissatisfied with the conventional accounts of the climate, and its ascription to the 'Central European type'. To the meteorologist, concerned with what we are beginning to call 'synoptic climatology', the North Italian Plain, mountain-girt but penetrable by 'tropical maritime air', presents peculiar if not unique conditions, which find any plausible parallel only in regions as distant as the Red Basin of China.

It is clear enough, then, that starting from the humble platform of the school synthesis we can diverge happily on specialist paths, but, if we do, when and how do we come together again?

' If seven maids with seven mops swept it for half-a-year,
' Do you suppose,' the Walrus said, ' that they could get it clear? '

And the Carpenter, represented even by some of our own number, still doubts it and sheds his bitter tear. But we are not all 'Carpenters' in this regard, and deeply conscious though we may be of the shortcomings of our present modes and methods we remain convinced that the effort at the higher synthesis must be made if geography is to survive and to thrive. Let us not dismiss the fuller contributions of systematic study as irrelevant or marginal detail. A few hours' study of the maps or a few weeks' residence in a region convicts the simpler pictures of gross unreality. 'There is nothing,' wrote H. G. Wells, 'so invigorating as a good generalisation, but it ought to go through its facts and marshal them ; it ought not to fly over their heads and expect them to follow.' But this, I fear, is what certain simpler modes of regional description quite frequently do.

My example, of course, merely sketches the problem but suggests both the nature of our difficulties and some possible directions of advance and improvement. The problem may be divided under the two heads of teaching and research.

3

In teaching regional geography it is better, in many respects, to travel hopefully than to arrive. What one is seeking to do is to convey the nature of geographical thinking. The emphasis is on the method rather than on the results. Little would be gained if the lecturer merely transmitted and the student duly repeated the lines of any one man's synthesis, however excellent. In these circumstances it is to be feared that, too often, the material passes from the lecturer's to the student's notebook without passing through the mind of either. What is required is a working demonstration of a geographer attempting a synthesis. In teaching, at least, it is surely incontestable that one man thinking as a geographer is better worth than any team of experts. It would, no doubt, be possible in the case of some chosen region, to produce in succession a geologist, a meteorologist, an historian, an economist, and so on, each contributing his own picture, leaving the synthesis to make itself in the mind of the student. Such a method would make great parade of scholarly thoroughness, at the cost of a complete denial of the geographical viewpoint. Syntheses do not make themselves in student or other minds and it would be ill practice to demand of students what we could not or would not attempt ourselves. If such self-actuating syntheses were ever to prove practicable, it could only be in the field of natural landscapes, uncomplicated by the developing hand of man. The succession, geology—geomorphology—climatology—pedology—plant ecology is, in large degree, coherent and serial : the inter-relationships move forward in an orderly manner like terms in a series. But the incoming of 'man the developer' transforms both the scene and the problem. It is a question now not of *processes* but of *actions*.

I believe that two vital corollaries arise from this fact. In the first place, it is idle and, indeed, fatuous to seek real analogies between the growth and development of human institutions or establishments and those, say, of land-forms. Certain of our younger historical geographers, observing how the geomorphologist reads his evidence and builds his succession from the actual ground, have become a little envious and proclaim their intention of trying to deal thus with, say, the growth of towns. But they are following a chimera and rejecting the better part of their own heritage. If 'angels of light', co-eternal with the cosmic process, had left actual records and maps, however imperfect, of what happened in the morphological past, the geomorphologist would naturally prefer them to his conjectures from field evidence. Though the instinct, even in historical geography, to get on to the ground is healthy enough, and will provide vital supplementary data, the laboratory of historical geography is the library and the muniment room, where the actions of man, deliberate, if sometimes wilful and irrational, are recorded. Towns do not grow, like rugose corals, by simple accretion, but rather by 'intussusception' ; they are reincarnated on their own ruins, and if their plan, in part, lives on, parts of their substance are altogether lost. It is no doubt a pretty student exercise to deduce the history of a town from the palimpsest of its architectural styles, but no clear truth could emerge without the dominating control of documents and maps. To force historical geography too closely into the mould of geomorphology is not to enhance human geography, but to underrate man.

The second corollary is equally important. It is that regional geography must sustain its rôle not only in the sparsely populated areas that are, methodologically, relatively easy to treat, but also in the 'climax' regions of old-established settlement in which the surface is fully taken up in the service of human economy. The principles of regional geography could not be based solely on the study of one or other of the southern continents or the emptier parts of Asia. The challenge lies in the areas of the great population groups, western Europe, eastern North America and southeast Asia. These differ not merely in degree, but fundamentally in kind, from the vastly more extensive areas of the world where the population density is less than 15 persons per square kilometre and in which man and his works are sporadic and generally 'recessive'. A vital aspect of the distinction is that in the one case the human background is history in its fullness, helpmeet and critic of the geographer ; in the other, to a large extent, it is anthropology, the science not of man but of primitive man.

It is necessarily in the 'climax' areas that the difficulties of our 'over-all synthesis' are greatest and its present weaknesses most apparent. Here we find a reason for the perennial, if friendly, disputation within the walls of the Royal Geographical Society between so-called 'academic' geography and exploration. To some of us the human geography of Somerset is more interesting and in many ways more significant than that of, say, Somaliland, and though we should wish both to be studied, it is the former, in general, which is neglected. The physical difficulties of doing so are admittedly less, but the intellectual difficulties are incomparably greater.

But even while our syntheses remain imperfect there is, of course, an impellent reason for widely ranging study of regional geography if only to obtain a reasonably full factual basis for a really comparative human or social geography. Like many others I learned from Mackinder the cardinal fact that, in respect of human development at least, each region is unique. 'China and Ind, Hellas and France, each have their own inheritance' or, as another poet in lighter vein has reminded us, our world,

> Holds a vast of various kinds of man
> And the wildest dreams of Kew are the facts of Khatmandhu
> And the crimes of Clapham chaste at Martaban.

The principles of physics can be learned within a single laboratory, those of geology adequately, if not completely, within the confines of our small island, but we must necessarily take the whole world for our province in seeking the principles, if such there be, of human geography. The specific character of geographical phenomena confers on our subject the quality that the methodologists call idiographic, and this is inevitably a great offence and stumbling-block to those seeking a strictly scientific approach, in the narrow sense. It might be, and indeed it has been, argued that, in the nature of the case, there can be no 'principles' of human geography, but I think the verdict is premature. But it can be only after much longer scrutiny and scholarly study of the world's specific regions that any such principles can emerge. Meanwhile I am sometimes moved by an unregenerate temptation to mirth and sardonic comment when I review present attempts at a 'generalised' human geography. Finch and Trewartha [1] inform us that :

> The American farmstead is composed of the residence, which is the nucleus, usually housing a single family, together with a collection of barns and sheds of various sizes and shapes, used for housing livestock and for storing crops and machinery. A fence commonly encloses the farmstead and separates it from adjoining fields.

Or again, concerning 'the distinguishing features of a city' :

> Within the downtown business district, tall closely spaced substantial buildings of brick, stone and concrete, occupied by retail shops and professional offices prevail. This emphatically is the hub or the nucleus ; upon it the street system converges so that within it the traffic is usually congested.

My quotation of these statements is not to be taken as a criticism of the authors whose courage and consistency in attempting what is undoubtedly the right *method* earns my admiration. The criticism, if such it be, lies against the present state of our subject and the absence of a thoroughly reviewed comparative basis upon which such statements can be made. I said that my temptation to poke fun at such statements was unregenerate ; in resisting it I remind myself that many an eminent palaeontologist owes his status entirely to faithful and methodical studies of the skeletal remains of colonial organisms such as the hydrozoa or the corals. It is far from evident that the 'exo-skeletons' of social man are less worthy of study. This at least is certain, that only by the methods of regional geography can we amass the data for a general human geography worthy of the name.

4

At this point we turn to the second aspect of my subject— research in regional geography. If what I have claimed is true, it surely follows that we should be active in the production of regional studies. Yet from the outset of our brief course as a higher discipline, this objective has proved very difficult of attainment, or at least has been rarely attained. So far as concerns regional studies of small areas it seems to be commonly

---

[1] *The Elements of Geography* (1st ed. McGraw, 1936), pp. 621, 633

assumed amongst us that they are little more than useful student exercises, fit subjects for graduate or diploma theses, to be decently interred in departmental storerooms. I dissent very strongly from this view, and I will go so far as to say that if many of those with whom I have been associated in examining such student theses had ever written one themselves, they would not only be better qualified to examine the theses, but could have made worth-while contributions to their subject. There are, however, obstacles confronting regional study, whether on the large or the small scale, notably our need of further analytical techniques. It is for that reason that I welcome the activity among our younger geographers upon select aspects of the geography of Britain ; these at least provide part of the material for regional syntheses. There is a growing tendency in some quarters to dismiss such careful local work as petty and parochial. I must protest emphatically against this foolish and unjustified attitude. There can be few of our younger geographers who would not eagerly grasp the opportunity of doing worth-while work abroad, but the facilities for such travel are very limited. Even if they were granted, we should not be absolved from the duty of studying our home regions. Nor are we at all likely to obtain the necessary support for wider travel by mere 'profession of faith'. Unless we give earnest of our methods by applying them in accessible ground for which information is full and where first-hand criticism is forthcoming, how shall we be listened to in studies less thoroughly conducted and less soundly based ? The cult, or rather the disease, of what I would call 'otherwheritis', is a real peril to the well-being of geography. At its best we may concede that it expresses a genuine and natural wanderlust, whether spiritual or sporting in its motive. At its worst it can only lead to a species of egotistic, impressionistic geography, a congeries of travellers' tales, agreeably titillating, no doubt, to the stifled imagination of strap-hanging town-dwellers, but making no claim whatever to scholarship. The observations and reflections, naïve or profound, of 'innocents abroad', whether explorers or merely travellers, will always, of course, be of great interest and value to geographers. There is doubtless a sense in which the diverse regions of our world exist in the minds and imaginations of the beholder rather than in objective fact, and the results, committed to paper, afford some 'fine confused feeding' and occasionally a few facts of real significance. But woe betide us if we seek to substitute such for careful, quantitative, and definitive work on accessible ground. 'What do they know of England who only England know ? ' still rings as a challenge to the geographer, but if by circumstance, certainly not by choice, we are at present [1] virtual prisoners in the land of our birth, there are worse places in which to perfect our tools, and outlook towers less central upon the life of man.

'But where,' we are asked, 'are the great magistral books on regional geography ? ' Leaving unnamed the small number we might timidly offer our outraged critics, the question is still one which requires a cooler, fairer, and more objective answer than it could obtain at the brief discussion at our meeting last year. Let us begin by admitting such share of the fault as doubtless lies in ourselves. Little wonder that we lack sometimes the rounded versatility required by our craft. There are still those amongst us who laugh off their almost complete ignorance of physiography or cartography as if it were some venial peculiarity like a weak back-hand at tennis or a slice at golf. Doubtless it is as little excusable in me that bus-route studies tend to bemuse me and that I should approach the study of 'urban fields' with the ham-handed ineptitude of the veriest tyro. The only honest answer to our critics is to ask for a little patience. Our predecessors in the departments were inevitably more occupied with the making of geographers than the making of geography. Under present pressure, we find ourselves in not dissimilar case, but with the important difference that now, as the half-century is about to strike, our departments, for the first time, are rather more adequately staffed. When our successors stand on shoulders as high and wide as those of the fashioners of the great traditions of English historical study, it will be time enough to expect the mature and balanced efforts of description and interpretation to which our theory points, if our practice fails to reach. Meanwhile, there is something on account to offer. Taking the case of our own country, and Mackinder's brilliant sketch, *Britain and the British Seas*,[2] as our point of departure, we note that there is hardly a chapter of that inspired and germinal book which has not been the subject of notable development and expansion in first-class specialist studies. I shall attempt no complete list here

---

[1] The address was delivered in 1951. [2] O.U.P., 1907

and propose no circle of bouquets, but I invite you to agree that the work of Professor Darby has illumined and will further illumine Mackinder's own distinctive field of historical geography, that Professor Stamp's great work on *The Land of Britain* [1] is rich and ample development of the very few pages devoted by Mackinder to agriculture, and that more recently Mr Wilfred Smith has repaid some of our debt to the applied economists and provided a real superstructure upon Mackinder's quite small beginnings. On the other hand, the unitary geomorphology of Britain is yet to be written ; I feel that I can safely promise that it will be written when I survey the very considerable efforts of our younger physiographers who maintain a good old tradition in British Geography, undeterred by disgruntled pedants, mis-claiming the fair name of 'humanist'.

That is my answer to the impatient and the importunate. We are on the job, by sound and tested methods, and we are not to be chivvied, bully-ragged, or even genially cozened into producing half-baked syntheses before we are ready. In his last and posthumous book, *The Idea of History*, R. G. Collingwood [2] has words that fit our case. He seeks to establish the true nature of history as the re-enactment of past experience, and he roundly condemns that 'scissors and paste history' which consists in borrowing and comparing excerpts from authorities. Sound historians, he says, study problems, they ask questions and, if they are good historians, they are questions which they can see their way to answer. Thus he vindicates Acton's precept, 'Study problems, not periods'. Yet no doubt when the questions have been asked and answered historians will come back to their periods. For us the comparable advice is, at least in the first instance, 'Study problems, not regions', but in general we shall be at least studying problems *in* regions and, if we are sound geographers, we shall be little likely to forget that the regional whole conditions our problems in numberless ways. Meanwhile we could, of course, write 'scissors and paste' syntheses, and well we know them—Structure, Relief, Climate, Vegetation, Rural life, Urban life, Agriculture, Industry, Communications, etc. Thus the compendium ; add them together and wait hopefully, Micawber-like, for something to turn up. But this, in the

[1] Longmans, Green, 1948   [2] O.U.P., 1946

language of symphony, is merely to announce our themes, to parade them seriatim. To the disgust and disillusionment of the better types of mind among our students, we use the method all too often in teaching ; it is fortunate that we have rarely had the hardihood to pretend it can avail in research.

5

And has the inspiring vision of regional synthesis now dis-limned and have I demolished the ideal which, at the outset, I professed to serve ? Not so, if I have succeeded in any measure in conveying my meaning. Our great objective remains, and as we work towards it, it governs our organisation and our methods. The high art will lie in the development of our simple, separate, themes, and the elaboration of a truly geographic harmony and counterpoint ; surely the geographical symphony, with its themes of Nature and Man, is characteristically and incurably contra-puntal. I will offer, in conclusion two practical suggestions by which harmonic wholeness may be assisted. Ideally, I believe, a geographical school should be organised in terms of regions, not of systematic disciplines ; that is to say, each member of the staff should commit himself to the study of a major region. Here I echo Roxby[1] when he wrote : 'The geographer's parish must indeed be the world, but it is too large a parish for all parts to be studied in detail by any one man. He must if he is entrusted with a university department, delegate responsibility for as many regional chapels-of-ease as he can find associates and colleagues to work them'. With the marginal *caveat* that I do not think the regional specialist will find much ease in his chapel, I cordially agree with this. And if some of you, as systematic specialists, find this an unwelcome doctrine you may well be impressed and perhaps persuaded by the fact that no less a systematist than W. M. Davis recommended a similar organisa-tion for the graduate school of geography at Clark University. For him the different divisions of systematic geography yielded 'only a discontinuous sort of knowledge' and in his view 'no-one should consider himself a geographer until he has become expert

[1] 'The Scope and Aims of Human Geography' *The Advancement of Science,* 1930, p.13

in the regional geography of at least one large area'. If this doctrine seems unduly to belittle our much-loved systematic specialisms, we can find evident comfort in the fact that your properly equipped 'regional man' must be, as it were, 'bachelor' of all the main contributory approaches and can profitably afford to be master of one. If, in this still early phase of the building of our subject, we find this doctrine too high and unattainable, I would qualify it with my second practical suggestion. Though in teaching the geography of a particular region any one man is better than two or more, in research some element of team-work is surely to be welcomed, In a 'regional' university, and all our universities in part at least are that, I think that a School of Geography can and should make use of the method of 'operational research', directing converging inquiries by different members towards that balanced final picture which it is their duty to produce. There are, it is true, obvious dangers in this method, but it will have at least the practical advantage of commanding some respect from sister subjects and lifting from the shoulders of the geographer the ungainly and overweening burden of an assumed omniscience.

For the end of our quest to see the regions of the world as wholes is not as yet. We stake our claim and point our objective, not as though we had already attained, but as those still pressing towards the high calling of geographer.

The Status of Geography and the Rôle of Field Work

MY TITLE associates two topics which have rarely, I think, been considered together, though I am sure that it is highly important that they should be so associated.

For a starting point I go back to what was in effect the inaugural meeting of the Association. In his reminiscences our first secretary, Mr B. B. Dickinson recalls how sixty-two years ago in January 1893 he distributed a circular calling a meeting of schoolmasters and others interested in geography, at Oxford. Mackinder took the chair at the meeting, and after summing up the discussion proposed the formation of an Association for the improvement of the status and teaching of geography. As I look round this morning it is instructive to recall that only two head-masters and representatives of eight schools attended ; six years were to elapse before the membership, now 3,700, reached the figure of 100. Even in our moods of frustration and disillusion-ment we can note, not indeed with complacency but with solid satisfaction, the progress since achieved.

As I look back to that founding motion of Mackinder's several reflections arise in my mind which I venture to pass on to you. Firstly I note that both school and university geographers were present and active at the inauguration, the latter in the person of its chief pioneer and exponent. I have watched with grief and consternation the widening gap in this Association between the school and university teachers of the subject. British geography is rather too prone to a certain fissiparous tendency. Firstly there is the cleavage, or difference in point of view, often manifest at the Royal Geographical Society between the 'explorer-cartographer' fraternity and the so-called 'academic' geographers. This is of old standing and need not concern us here except to note that in this as in other cases 'academic' is an adjective applied to those who understand their work by those who do not. But let me say very flatly that this Association will ill advance the twin aims announced at its foundation if the two main bodies of teachers of the subject do not remain in close contact and frequent conference.

The Joint School of Geography at King's College and the London School of Economics is in some sense the host department for this January Conference, and I have myself attended it regularly since my first appointment as a junior lecturer in January, 1922. Yet during this period of thirty-three years I am only the third university professor, in office, to have been honoured by election to your presidential chair. Of far more import is the, for me, grievous fact that this year, for the first time, the dates of the conference are such as to make it almost impossible for university geographers to attend. The Institute of British Geographers is concurrently in session at Durham, a meeting which my colleagues and I are naturally keen to attend, for the Institute is essentially the representative and meeting-place of the university geographers of Britain. If we want effectively to impinge upon or 'present a front to' the world at large or the educational authorities, we should do well to adopt as marching orders the familiar words of the hymn 'we are not divided, all one body we' and to make it so at least to the rudimentary extent of planning and controlling the dates of our meetings. I am not unaware of the difficulties but I do not believe they are insuperable.

For many years following its foundation, twenty years ago, the Institute of British Geographers functioned in some sense as the university section of the Association. But with its vigorous present life and a membership of 400, it has been natural, inevitable, and no doubt desirable, that the younger body should move towards full independence. But it is not beyond the wit of man to secure a sufficient measure of spatial and temporal juxta-position between the two conferences to render cross-fertilisation possible.

But you may be moved to ask and you will in fact be wise to ask what forces or tendencies foster the separation I have noted. My own diagnosis is as follows. On the one hand the university teachers are inclined to feel that our programmes are too preoccupied with 'education' and too little with geography or, to be more explicit, what may be called 'research geography'. On the other hand, when we do publish papers which are additions to knowledge I am told of outbursts from teachers who say that they are not interested in research. I must beware here of the invidious pitfalls of picking examples, but I will risk one. Turn back to *Geography* for April, 1953. This is the sort of part which warms my own heart since it registers advances of geography as a subject. It includes *inter alia* Mr Waters' paper on 'Aits and breaks of slope on Dartmoor steams'. But when I ventured to express to our Honorary Secretary my appreciation of this part and this paper I was told that I should have seen the adverse comments in letters from primary school teachers and others. We will let them remain anonymous but I put it now to you personally : Is your attitude the narrow and, I do not hesitate to add, contemptible one that you have no use for it because there will be no questions in the G.C.E. examination on aits or Dartmoor streams? If so I charge you with being in the worst sense 'non-academic' geographers, conspicuous for ignorance and lack of imagination. I cannot develop the theme fully here, but consider only for a moment what river islands have meant in terms of river-crossings and town sites. Do we then accept them as inexplicable 'Acts of God' and despise careful investigation designed to illumine their origin? Is this as some might hasten to say 'mere physiography'? If these are our attitudes we shall ill-advance the status of our subject and I doubt also if we shall be on the road to teaching it better. In my view it is not the physiography that is 'mere' but the geographers who merit this term in the sense that they are 'hardly' geographers.

If the Association were to turn its back on research geography, there would indeed be small hope of engaging and retaining the interest and support of the university geographers. In justice let it be said that it was not the policy of Professor Fleure so to exclude it. He maintained an admirable balance in our journal. Moreover, since our present Honorary Editor brings, in my judgment, one of the finest research minds in geography in this country to his work I do not expect that research papers will in fact be excluded. We need not question of course that some such are more suitable and of more general interest than others and that our editor and editorial board have an incessant and onerous responsibility for choice. But never let us forget that if it is the status of our subject we would enhance, then the doubt we have to meet is the one so widely held, whether we are in fact a subject in which real research—real widening of the bounds of knowledge —is possible outside the field of direct exploration.

But if I speak bluntly or even a little harshly of some of those who are not in my judgment furtherers of the status of our subject, let me in justice turn for a moment to the other versant of the dispute. To my university colleagues I would say that we owe the plainest of duties to support the Association for two good reasons. Firstly because our grammar school colleagues lay the foundations on which we build, and secondly because, if we have any sense either of gratitude or historical perspective, we recognise in the Association the foster-mother of our own highly valued Institute. As Secretary of the Inaugural Committee of the latter I take leave to state that without the Geographical Association there would have been no Institute of British Geographers. The relationship of affiliation is as real as that of our own Association to the Royal Geographical Society.

But the difficulty at which we have glanced is, I believe, merely an aspect of a wider problem, that of the educational relations of our subject. Since in what I am about to say I run a grave risk of misrepresentation may I preface my remarks by affirming that I have no doubt at all about the vital educational rôle of geography and the need of having it taught and well taught. If I sought in brief words to define our aim here, I should borrow a favourite phrase of my old chief, Professor Rodwell Jones, and say that it is our rôle to inculcate 'an awareness of environment'. I must not forget here the American colleague who informed me that he did not know what I meant by environment. I suspect that he was desiring to convict me of what has been regarded as a heresy in the view of orthodox American geography under the name of 'environmentalism'. I do not plead guilty to this charge, and in case these words should meet the eye of the gentleman in question I will say only that if he would care for a few months to associate himself with the Joint School of Geography, we will teach him very quickly and unarguably what environment means.

I have no doubt, too, that the training given in the University Schools of Geography is as good an education as is given by any other subject, and better than most.

But I am disturbed at the contest between matter and method

which goes on in training colleges and training departments. Summoning my not inconsiderable resources of rashness, I will here and now accuse the 'education people' of having been potent agents in diminishing the status of our subject and retarding its advance. It was after a depressing conversation with two training-college teachers at this conference that some few years ago I spoke at our own Institute of Education under the jeremiad title 'Geography ; is there any hope ?' My theme was that the training colleges and departments were too little interested in geography itself in its full developing depth and scope, and too preoccupied simply with methods of teaching it. I readily conceded to my friend Mr Honeybone that the two things are not incompatible, but I remain persuaded that a proper balance between them is often not kept.

This really takes us back to Mackinder's inaugural motion which I have quoted. Had I been present at that meeting with the views I have since developed I should have desired to point out that the improvement of the status and the teaching of our subject are two quite distinct though not unrelated tasks. No refinement of teaching artifice, no compilation, however shrewd, of tips and methods, will of itself convince the world of learning that we are a full and dignified discipline. For this we must look to the matter of our subject and its research progress. Do I, perhaps, exaggerate the antithesis here ? If so it is at least a view based upon my own experience. I remember for instance my anger and mortification at being summoned to a Ministry course at Oxford to talk to teachers about the philosophy of geography, only to hear my class told before I had uttered a word that it really didn't matter very much what geography was or might be, since their task was *not to teach geography but to teach 'George'*. Thus was alliteration's artful aid applied in support of that most dangerous, damnable form of heresy which is based upon a half truth. As one who is proud to have been a student of Sir John Adams, Sir Percy Nunn, and James Fairgrieve, this amazing announcement so far as it contained any truth was not news to me. May I add that I am not willing to accept the customary gibe that my views are prejudiced by the inevitable outlook of a university professor. That is the customary evasion of educationists when I seek to argue with them. May I therefore place on record that my first teaching of geography was to the sixth form in my own school and that I served a respectably long apprenticeship teaching in primary schools in both north and south London before my first appointment at King's College. And I remain to this day closely conversant with the problems and possibilities of grammar-school teaching through the work of my wife.

It is in the light of such experience and knowledge, as well as that derived from my present full-time occupation of training geographers, that I voice my grudge against the training colleges and departments for failing too often to maintain the work of the universities in clarifying and strengthening the aims and claims of geography as a subject. Let me refer to one other actual incident which has contributed to colouring my views. I was invited not long ago to assist in the selection of a lecturer in geography at a well-known training college. The chairman of the interviewing board was a mature historical scholar. But he and I were unable to educe from most of the candidates any evidence that they were still very interested in the subject in which they graduated. They were manifestly all too interested in various rather cranky ideas and devices, visual or other, for the teaching of the subject. Most of them indeed were men or women with a mission with some brand-new teaching method to propagate. I was reminded of the remark of Canon Raven that psychology seems to obtain too much of its data from infants and invalids. I am happy to say that we appointed to the post the only man in the group who appeared still to be a geographer concerned with the advancement of his subject, rather than an amateur educational theorist.

The best summary of the position was contributed recently in conversation by my friend and old student Mr C. L. Heywood, who remarked that the greatest obstacle to the better teaching of geography is ignorance of geography, i.e. the matter of the subject ; not be it noted of 'George' or of child psychology. In justice I note that this comment comes from the educational side of the fence, since Mr Heywood is a Lecturer in Education.

Having reached this point of my address I paused to read what I had written and it seemed to me that I had so far adopted a rather complaining and querulous tone. I recalled that my friend and colleague Professor R. O. Buchanan had contributed to *Geography* a paper entitled 'Some Grumbles', and I began to fear that this address might be more appropriately entitled 'Some More Grumbles' and placed under the *nom de plume* 'Gama' from Gilbert's *Princess Ida*. At this point therefore I seek deliberately to change the key, turning to the second part of my subject and attempting a more constructive approach. It will enable me also to rebut the possible charge of being inexcusably prejudiced against the study of teaching methods as such. I remain keenly and actively interested in field work methods in geography, and I will even venture here to lay down the broad principles which I think should inform field teaching. I want here indeed to try to state as simply and definitively as possible what I conceive to be the proper aims and therefore methods in the teaching of field work in geography. To this extent at least, as I here avow, I am prepared to enter the field of the teaching method 'cranks', and if the qualification for 'crankhood' is to hold strong views on the matter I am both willing and able to join issue here.

All geographers no doubt acknowledge, in the sense of 'pay lip service to', the importance of field work in their subject, but they show considerable doubt as to its objective and methods and pay far too little attention to it. This is partly no doubt because it is difficult to organise and conduct, but I think it is also true that the wide scope of our subject tends to confuse the issue. To note the divergence of aim and method one has only to compare the practice of our now numerous University Schools of Geography on their regular annual field classes. In some cases this is little more than a sight-seeing tour, in others some form of detailed local investigation is attempted. For a few, field work in geography constitutes surveying and nothing more. We must in any case clear the ground by agreeing that field work cannot simply be equated with the sum-total of extra-mural activities. Visits to works, farms, etc. are not in fact field work at all in the sense in which I am here advocating it. I need hardly say that they have their evident place and value. My objection is that they so often take the place of real field work, since they offer an easy and lazy way out of a difficulty. Other people, commonly unpractised in the art, attempt to do the teacher's work for him, using a maladroit lecture method against a background of industrial or rural noise. It goes without saying that information thus gained, skilfully interpreted and placed in its proper context can be of high geographical interest.

In general, however, I submit that the object of field teaching, at least in the elementary stage, is to develop 'an eye for country' —i.e. to build up the power to read a piece of country. This is distinct from, though plainly not unrelated to, 'map-reading'. The fundamental principle is that the ground not the map is the primary document, in the sense in which historians use that term. From this first principle I pass to a second, that the essence of training in geographical field work is the comparison of the ground with the map, recognising that the latter, at its best, is a very partial and imperfect picture of the ground, leaving it as our chief stimulus to observe the wide range of phenomena which the map ignores or at which it barely hints. In Britain we are admittedly favoured with an exceptionally full equipment of maps. We can generally bring together and compare at least the topographical, geological, and land-use maps. To these are now being added the sheets of the Soil Survey. But even when all such material is synthesised it remains true that the ground will reveal innumerable significant details which the maps ignore.

This second principle carries a corollary so important that it is almost a third principle, namely, that the order of working is from the ground to the map and not vice-versa. It is rank bad teaching, to my view, to let a class occupying a view-point bend their heads over their maps, instead of concentrating on the work of looking and seeing. Let us say with Wordsworth :

> Enough of Science and of Art ;
> Close up those barren leaves ;
> Come forth, and bring with you a heart
> That watches and receives.

When leisurely observation is complete comparison with the map can fitly proceed. The reverse order too readily evokes the stupid and erroneous comment : 'Well what is the point of all this ? It's all on the map anyway !'

I need hardly add that it is inherent in the method and point of view I am commending that when in the field one should concentrate on observable field data. An easier but grossly incompetent method for the ignorant or unpractised is to deliver a roadside lecture of guide-book or text-book information borrowing freely but irrelevantly from history, geology, or agriculture. The question is rather what can you see from here and more particularly what can you see which the map fails to portray. From this we can lead straight on to my last principle that the essential 'doing part' of geographical field work lies in making significant additions to the map. This should be placatory to those of you who in conformity with one of the popular educational lunacies of our day are clamouring for 'Activity'. Much of what is urged and taught in this connection appears to me ineffably silly, patently erroneous, and intellectually beneath contempt, but there is evidently here a place for *doing*, and it is a first-class exercise to 'walk' a 1 : 25000 sheet annotating and interpreting it on the ground, making, that is, additions to the map. Broadly speaking I think these additions are at their best where, if they are not directly cartographical, they are at least graphical. To take one simple instance, it is a highly instructive and rewarding exercise to map a series of river terraces by putting on the map the breaks of slope which bound them. Nor is the exercise 'mere physiography'. In lowland country, the lines drawn will be closely related to the land-use pattern. In upland country this will still be true and in addition significant light will be thrown on the siting of settlements.

I must not here dilate at length on possible exercises of this kind, but I have in fact done a better thing by persuading my friend Mr Hutchings to exhibit at our Conference some of the exercises which he uses with school parties at Juniper Hall.

May I now turn back to the wider field of argument in which field work must be vindicated and reply to some possible objections. You may perhaps lament that the sort of work I have indicated is narrow and 'local', implying that for that reason it is in some sense trivial. To this I reply with G. K. Chesterton that to make a thing real you *must* make it local, and I am completely persuaded that geography begins at home. What we have to develop if we seek status for our subject is the art of seeing and using accessible local ground as a laboratory for our teaching. Though I flatly and forcibly reject the charge that what is local is trivial I can follow the mind of the objector ; he is craving for our valuable and characteristic 'world view,' he is one of the geographical 'Wesleys' who is claiming the whole world as his parish. This is well enough providing only that he does not try to run before he can walk. The test of a geographer young or old is, in my judgment, not whether he is topically up to date in figures of world production, but whether he can interpret the full interest and significance of the 'Little Lands' in Belloc's sense.

> The thousand little lands within one little land that lie
> Where Severn seeks the sunset isles or Sussex scales the sky.

Or as Professor East and I have ventured to remind our readers in the words of Blake :

> Nature and Art in this together suit :
> What is most grand is always most minute.

It is here that we reach the core of the argument I am seeking to present. The road to the attainment of both of our objectives, the improvement of our status as a subject and our teaching of it, lies in the development of the laboratory spirit and the careful, indeed minute, study of limited areas. I am well aware that this conclusion will be repugnant to those who are primarily concerned with the study of large and distant divisions of the world, such as the Far East and the inter-tropical belt. The difficult piece of integration which geography attempts must be vindicated on ground more accessible and manageable than this. I would recall that much of the best work of our pioneers in establishing the subject and training its teachers was by way of the Oxford and London diplomas in the subject with their characteristic and valuable dissertation on a small area. Those Honours schools in this country which retain this method have something of the highest value which I hope they will not readily let go. It is in and by such studies that geographers are trained, not by encouragement to absorb and repeat the professors' dicta in the nightmare world of 'geopolitics' or 'world economic geography'.

It is perfectly manifest, of course, that in teaching we have to solve the problem of balance between local and wider aspects, and the high art is to use field work to illumine the wider field. This can very readily be done, notably in the field of physical geography. W. M. Davis quoted more than once these significant and forceful words from the preface of Huxley's *Physiography* :

I endeavoured to give them a view of the place in nature of a particular district of England—the basin of the Thames—and to leave upon their minds the impression that the muddy waters of our metropolitan river, the hills between which it flows, the breezes that blow over it are not isolated phenomena to be taken as understood because they are familiar . . . to show that the application of the plainest and simplest processes of reasoning to any one of these phenomena suffices to show lying behind it a cause which again suggests another until step by step the conviction dawns upon the learner that to attain even an elementary conception of what goes on in his own parish, he must know something about the universe.

There in clear and authoritative words is indicated the royal road by which careful local study can be projected as a search-light beam into our wider universe of discourse.

And here too, unless I am much mistaken, lies the clue to the still unanswered problem, the devising of suitable geography syllabuses for the Secondary Modern schools. I should require these to be planted firmly and centrally upon local study.

But the 'projection' of which I speak is not confined to physical geography. There are endless opportunities for applying the method of 'comparative regional geography'. For instance, a day in the central Weald prompts a comparison not as might at first appear with the formal structural analogue, the Boulonnais, but if we consider soil, vegetation, and human settlement, rather with the Argonne. In the same way South Wales affords us a comparison for Pennsylvania and the south Pennine area for the 'Newer Appalachians'. All this and more can arise easily and naturally from field work done in Britain, providing it is done slowly, carefully, and on foot, on the lines I have already sought to indicate.

I am less enamoured of the practice of dashing around Britain or across Europe by charabanc. Before this can profitably be done, eye and mind must first be trained by field work of laboratory intensity. Otherwise the method too easily justifies Mephistopheles' sneer in Goethe's *Faust*

> Are Britons here? They travel far to trace
> Renowned battlefields and waterfalls.

In conclusion may I also call attention to the opportunities for widening the bases of school field work by using the four centres of the Council for the Promotion of Field Studies. Juniper Hall,

76

near Dorking, is our chief centre of geographical teaching by virtue of the signal laboratory facilities offered by the classic Surrey hills and of the genius as a teacher of our Warden, Mr Geoffrey Hutchings.

The other three centres greatly extend the regional range. Malham Tarn House gives an introduction to the geography of the upland zone of Britain. Flatford Mill avails for an introduction to our distinctive eastern province and prompts many revealing analogues with the north European plain, in Germany and Poland. Finally, Dale Fort in Pembrokeshire introduces the ria and plateau coast of the Atlantic zone of Britain and affords striking examples on accessible ground of the landscape imprint of two contrasted cultures—Celtic and Anglo-Saxon.

Speaking as the senior honorary executive officer of the Council I need hardly emphasise to this audience of teachers the advantages of ground already prepared and studied by our resident staff.

I am perfectly well aware that the greatest single obstacle to field teaching in geography is the shrinking of the teacher from his or her own ignorance of unknown country. All this the method of the Council for the Promotion of Field Studies is designed to cure. When my friend Mr Francis Butler started the Council and founded our four centres he took, I consider, the most important step since the founding of our own Association, for the improvement of the status and teaching of geography. No grammar school boy or girl should complete his course without a visit to each of these centres. It is our aim and intention to develop the geographical side of our work at Malham, Flatford, and Dale, to the Juniper Hall standard, and in guiding the educational policy of the Council I have proposed to the Ministry of Education that we should expand our activities so as to bring in the Secondary Modern Schools.

At these centres you can consolidate your teaching under proper class-room conditions with the necessary equipment and expert local guidance provided. And if your work becomes linked or tinged with the Council's other main interest—biology—so much the better, in my view, for your geography.

You will not, I hope and believe, want to interject the false and futile ' groan' that all this is not examination work. You would be partially wrong even in fact, but even more so in spirit.

77

I have ventured already to quote the great Nature poet Words-worth in support of my thesis. I will go further and essay a slight parody of famous lines thus :

> One traverse in a Surrey Vale
> (or if you prefer it Yorkshire Dale)
> Will teach you more of Man,
> Of Man in his terrestrial home,
> Than all the text-books can !

POSTSCRIPT

A typescript copy of this address was sent by request to *The Times Educational Supplement*. The extracts which appeared in that journal taken by themselves might suggest that I was chiefly concerned to make an indiscriminate condemnation of the work of training colleges and departments. Such indiscriminate condemnation would have been wholly unjustified for two good reasons. In the first place no university teacher could profess a sufficiently wide knowledge of such colleges and departments. But further, there is not the slightest doubt that a great deal of excellent work has been done in the field of teaching method in geography. I have no doubt for instance that it is much better taught at present than history. Let me say therefore with vigour and emphasis that I have no feeling of criticism and condemnation directed at teachers of 'method' in general. And I should wish to add that within the University of London I have always been conscious of substantial concurrence in aims and emphasis with my colleagues of the Institute of Education, King's College, and at Goldsmiths College. It is no less than a duty, too, to acknow-ledge also the excellent work done in the Colleges of the former Group II of the London Training College scheme (i.e. Furzedown, Stockwell, and Whitelands Colleges). These, for many years, were the King's College Group and were known to me as their Group examiner. May I also pay tribute to the excellent work done in Environmental Studies by Southlands College.

If I were now asked how, in the light of the favourable testimony I here acknowledge, I can justify the criticisms I have made or implied, I should reply that they are based upon the experience of my own students. Each session twenty to thirty honours graduates in geography leave us to complete their training as teachers in a wide range of departments. It would be unwise to take student reaction too simply at its face value, but I cannot completely dismiss or discount the widespread frustration and disappointment which many of them feel as 'method' progressively forces them away from the subject they have learned to love and respect, and in which their intellectual capital is naturally and properly vested.

6

The Rôle and Relations of Geomorphology

IN CHOOSING a subject for this lecture I have been somewhat embarrassed by the fact that, not many years since, I delivered another such lecture at a sister college in our federal University. On that occasion it seemed appropriate to deal with certain aspects of the philosophy of geography in the large—a subject rich in differences of opinion and, therefore, of perennial interest to geographers. In the years between, I might perchance have changed my mind on this subject ; there is at least ample warrant in example for such fluidity in opinion. Alternatively, since no-one has formally attacked the propositions which I then presented, I might, in a truly Greek spirit, seize this occasion to answer my own pronouncement, pointing out its evident heresies and short-comings. But though the latter are not in doubt, I have not in fact changed my mind and conclude that I cannot profitably return to the topic. I prefer rather to pass from the whole to the part, from geography as a compendious discipline to geomorphology, that special division of the field which has engaged a great part of my own interest.

There is, indeed, another good reason for my choice of subject. This session marks the departmental separation at King's College of geography from geology, with which it had been grouped for a quarter of a century. As a member of the staff of the old double department, I inevitably inherit its traditions. I am not ashamed of my geological parentage, nor in any way inclined to sever the intellectual liaison between the two earth sciences. That liaison is most clearly seen in the field of geomorphology.

Geomorphology has, naturally indeed, been regarded as a border-line study falling between the fields of geology and geography. Its rôle and relations, however, have been the subject of some doubt and dispute. Now that geography has attained full recognition in the universities beside geology, well established much earlier, it is desirable that the real nature and contribution of geomorphology should be more closely examined.

It has, of course, never been doubted that the roots of geomorphology lie and must always lie in geology, but the story of its emergence as a co-ordinated subject is worthy of a much closer historical scrutiny than it has yet received. The great nineteenth-century age of geology has now receded sufficiently in perspective to make possible adequate historical study of the growth of ideas. Britain played a cardinal rôle in the fashioning of these ideas and the views of British geologists are thus highly significant. In any case, a geomorphologist may be forgiven for believing that, in the sociology of knowledge as in the study of land-forms, the past is the key to the present.

We may start our inquiry by noting that Humboldt and Ritter, the founders of modern geography, though familiar with the geology of their days, were necessarily inheritors of the traditions of Werner and the Freiburg School. This afforded no evolutionary viewpoint in respect of the face of the earth ; its attitude was static and 'structural'. It was Hutton and Playfair who sowed the seed of geomorphology, but though Playfair's *Illustrations of the Huttonian Theory* was published five years before Humboldt and Ritter first met as young men in Berlin, there is little evidence that they ever learned the geographic force of the Huttonian doctrine. Their geology was, indeed, not only pre-Huttonian, but pre-Lyellian, though this is less significant than it might, at first sight, appear. There has been too great a tendency to regard Lyell as the literal disciple of Hutton. So far as concerns geomorphology their points of view differed radically. The ideas of the 'normal cycle of erosion' are implicit in the pages of Hutton and Playfair, but Lyell long retained something of the 'structuralist' point of view in regard to valleys. As late as 1838 we find him treating the Wealden valleys as essentially fissures, slightly modified by sub-aerial action, and he was certainly no whole-hearted convert to 'erosionist' views during those critical years in which geology began to take formal philosophical shape. The reality of the antithesis between the viewpoints of Lyell and Hutton is shown by the publication, as late as 1857, of George Greenwood's *Rain and Rivers*, with the significant sub-title, *Hutton and Playfair against Lyell and all comers*. It is difficult to realise today that the efficacy of sub-aerial denudation, and particularly of river erosion, remained in dispute in Britain until the years 1860–70. These years saw the publication of the work of Jukes on the rivers of Southern Ireland,[1] and Whitaker's paper on cliffs and escarpments.[2]   The latter, even at so late a date, contained a doctrine too radical for the Geological Society and was published in the *Geological Magazine*.   Topley's great memoir on the Weald came early in the following decade, and succeeding years witnessed the labours of Dutton, Powell, and Gilbert in America, which all are agreed in regarding as the mainspring of later thought in the field of land-sculpture.

It was well-nigh quarter of a century before these views became fully current in Britain.   The well-known papers by W. M. Davis on 'The Development of certain English Rivers'[3] and 'The Drainage of Cuestas'[4] marked the appearance of the new mode of thought in British journals.   Nor were British geologists wholly unreceptive of the methods and their results.   Marr's *Scientific Study of Scenery* appeared in 1900, and Archibald Geikie's great textbook included a final twenty-five pages on physiographical geology, which, though it is little more than an appendix to a major work, gives adequate references to the whole field as then known.

It was at this stage that the first signs appeared of a desire to relegate geomorphology to geography, or at least to treat it as a borderline field.   Marr's preface refers to geomorphology as 'a subject which has sprung from the union of geology and geography', and Geikie's opening paragraph on 'Physiographical Geology' concludes with the words : 'The rocks and their contents form one subject of study, the history of their present scenery, another'.   Equally significant is James Geikie's well-known work on *Earth Sculpture* (1898).   This gives adequate attention to denudation in the large, but retains a certain emphasis derived from older 'structuralist' views.   It gives a condensed and somewhat misleading account of river development, but, in the preface, it is made clear that 'geographical evolution', a subject 'studied with assiduity by Professor W. M. Davis and others in North America', does not come within the compass of the book.   We may note also that H. B. Woodward's *History of Geology* (1911), which necessarily reflects the point of view of British nineteenth-century writers, includes only three index references to the word

---

[1] *Quart. Journ. Geol. Soc.* **18** (1862), 378
[2] *Geol. Mag.* **4** (1867), 474
[3] *Geog. Journ.* (1895), 128
[4] *Proc. Geol. Assoc.* **15** (1901), 75

'denudation', and refers to 'the fascinating subject of earth sculpture' only in its concluding remarks and then only to note that 'the subject is by no means a wholly geographical one', since fragments of ancient landscapes are occasionally revealed 'fossil', as beneath the Torridon Sandstone or the Trias at Charnwood Forest.

In all this and much more that might be quoted there is an implied distinction between the study of land-forms and geology proper. It is evident that some British geologists, at least, thought of land-form study as 'geography'. Nor were they alone in this view, for Mackinder has admitted that in his earlier days he was attracted by Geikie's antithesis and was inclined to regard the laying down of the rocks as geology and their shaping into a surface as geography. Yet we note that this is not what Geikie said ; his grip of his subject was far too sound to permit any such exclusion of land-sculpture from geology ; it was physiographical geology. Nor need we suppose that all other geologists wished for such exclusion. There was, indeed, no systematic or organised subject of geography to which they could relegate this topic ; it was an intra- not an inter-departmental partition which they were seeking to erect. No useful purpose whatever is served by criticising their attitude on grounds of logic ; it is better to inquire why and how the situation arose. Our conclusion must be that it arose from the necessary character of geology itself combined with an historical accident.

Geology in Britain perfectly exemplifies the principle expressed by Mill when he pointed out that the definition of a science 'like the wall of a city . . . has usually been erected not to be a receptacle for such edifices which might afterwards spring up, but to circumscribe an aggregate already in existence'.[1] The study of land-sculpture, beyond a very broad and elementary stage, was doubtless excluded from formal geological instruction in earlier years because it was a disputable topic. This and the late closure of the controversy constituted the historical accident. If Hutton had written less obscurely and Playfair's *Illustrations* been more widely known, events might have taken a different course. By the time that geomorphology attained to vigorous growth, geology had fashioned a full curriculum of formal training.

---

[1] *Unsettled Questions of Political Economy* (1874) p. 120

The Geographer as Scientist

Archibald Geikie's textbook, representing the Edinburgh courses, and the regimen laid down by J. W. Judd at South Kensington, which became the basis of the first London degree in geology, sufficiently indicate both the character and the problems of such formal training. If stratigraphical or historical geology be the culmination of the subject it must necessarily rest upon the prior disciplines concerning the materials and processes of earth-history. Petrology carries its necessary prelude of mineralogy and crystallography. Palaeontology must be, if necessary, an *ad hoc* study, with or without some measure of biological background. Physical geology is the forerunner of stratigraphy ; it can be, and perhaps must be, in large part a study of rock-making. The traditional geological curriculum necessarily ranges these separate and special studies in series rather than in parallel ; there is a long preparatory apprenticeship to be served before the fullness of geology is attained, and this is to say nothing of the ancillary scientific disciplines. To some it has, no doubt, always seemed, with H. G. Wells, that geology is, for these reasons, 'an ill-assembled subject'.[1] But it is difficult to see what alternative method could have been followed by its early teachers. What here concerns us most is, that in this scheme, geomorphology must figure either at an early or a late stage. In an early stage it forms an admirable introductory vehicle in training, and this is where it has generally been placed in courses. As a field for research it comes after stratigraphical geology in logical order, since it is answerable for the last brief chapters of the geological record. From this point of view Geikie's treatment is wholly sound ; physiographical geology completes and complements the stratigraphical story. Yet it can be omitted from the story without loss to its epic sweep ; it is little more than an appendix to the whole.

There are several aspects of this fact at which we may look more carefully and which go far to explain and even, in part, to justify the attitude of some geologists to geomorphology. Land-forms, as such, necessarily play a relatively small part in geological history as a whole. The present marks a mere stage in the geologist's series of superimposed pictures. Though similar

[1] *Experiment in Autobiography*, Vol. I (Gollancz and Cresset Press Ltd, 1934), p. 226

diversified landscapes have existed before, of them 'nothing stands'; their hills have indeed been shadows which have gone beyond recall, save where chance fragments have been buried and exhumed. There is nothing that the geologist can really compare with the present physical landscapes of the earth, and a close scrutiny of them throws comparatively little direct light on general geological history. In the narrower sense, at least, geology is chiefly an affair of sea-bottoms not of land surfaces, even when the full tale of the great continental formations is told. The substance of geological science is locked up in the cumulative pile of sedimentation and the records it embodies, including those of earth-movement. If the mind of man had been brought to bear at a later stage of the geological cycle when the continents were far advanced towards base-levelling, the progress of geology might well have been hindered by paucity of natural exposures, but it would still have been possible from the evidence of outcrops and borings, supplemented perhaps by geophysical methods, to discern the structure of the crust and reconstruct its history. But for geomorphology there would have been little scope, save for the fact that the nature and reality of peneplains might then have been all too plainly evident ! Such considerations as these must obviously and rightly affect the amount of attention which the geologist gives to land-sculpture as such, within his crowded programme of study. They undoubtedly led to some over-emphasis of deposition at the expense of denudation in the older geology. Hutton was clear that the two are related like the halves of an hour-glass, but not all the nineteenth-century geologists had views equally sound. The point is an elementary one, yet there is no doubt that the teaching of stratigraphical geology was for long unbalanced by an imperfect view of the source of sediments. The tendency is expressed to the point of caricature by a quite ludicrous diagram which Professor Hogben has allowed to appear in his *Science for the Citizen*. It is entitled 'Denudation and Deposition' and shows first a group of mountains rising above a coastal plain, next the submergence of the coastal plain and the unconformable accumulation of a marine cover upon it, and finally its re-emergence. Throughout the whole of the period of this considerable episode of deposition the mountains of the background stand immobile and unaffected in a single feature of their outline by denudation ! It is needless to say that no geologist worthy of the name could perpetrate such an error as this, yet it does represent, in some sense, the curious lack of balance between knowledge of the two co-ordinate phases of the geological cycle. Deposition is self-registering, denudation, so it has been thought, is not.

The last point brings us at once to another question of importance responsible for a certain amount of misunderstanding between American geomorphologists and European geologists. By the end of the nineteenth century the stratigraphy of western Europe, and Britain in particular, was known in far greater detail than that of North America. The physical history or 'palaeogeography' of the British Isles had been written by Jukes-Browne with a completeness not then attainable for any other region in the world. To form an acceptable sequel to such a history it was necessary that British geomorphology should be 'chronological', The system of geomorphology propounded by W. M. Davis *was* in a sense chronological or sequential, but it could not be grafted on to a stratigraphical system, nor was it designed for such a purpose. The cycle of erosion was of indeterminate length and dealt in stages, not ages, as Davis always insisted. It enabled one to sketch in broad terms the physical history of regions of which the stratigraphical history was unknown or imperfectly known, and hence it was and is a powerful tool in reconnaissance. But Britain and much of western Europe did not need reconnaissance. It is interesting to note some of the first applications of Davis's work by British geologists. Cowper Reed, in the Sedgewick Prize Essay for 1900, dealt with the rivers of eastern Yorkshire, dividing their history into six cycles. It was impossible then and is still impossible to relate this cyclical chronology to the stages of Tertiary stratigraphical history. On the other hand A. E. Salter [1] saw in Davis's theories a means by which he sought to clarify the tangled history of the superficial deposits of the London Basin. He classified the various deposits in 'drift series' arranged in descending order, each marking a stage in the history of a consequent stream entering the Basin by one of its marginal gaps. This classification cut right across the stratigraphical grouping, which treated the deposits as marking various stages of the Pliocene or Pleistocene periods. Davis him-

[1] *Proc. Geol. Assoc.* **19** (1905), 1

self put forward his 'two-cycle theory' of the evolution of the southern British drainage.[1] He ignored the evidence of the marine Lenham Beds (L. Pliocene), finding evidences of two successive cycles of sub-aerial denudation. Subsequent work has shown that for a great part of the area this theory is completely in error. The Lenham Beds cannot be ignored, and from our knowledge of their former extent we can now perceive that the drainage of a great part of the area is superimposed from a sheet of Pliocene deposits and is in its first cycle. But the basic evidence for this conclusion is stratigraphical. In brief, one may say that British stratigraphy was too well known to render it possible or useful to think and speak in terms of cycles. The geologist tended to feel that, where he had any evidence at all, it was both more tangible and more precise than that offered by morphology alone. From another standpoint it could, of course, be said that the stratigraphy was not well-enough known to afford any chance of correlation with morphological cycles. What is wanted as the complement of the stratigraphical story is the denudation chronology of Tertiary times. The Tertiary deposits are limited in area, small in thickness and concentrated almost entirely in the South-East Province. Over much of the British Isles even the Chalk is missing as a datum formation. In the west and north of the country stratigraphical perspective is thus inevitably lost. The summit-plane of Wales has been variously identified as an exhumed sub-Triassic or sub-Cretaceous surface and as a Tertiary peneplain, and there is no direct geological evidence, in the narrower sense, to decide between the alternatives. The demonstration, in later years, throughout Britain of a large number of planation surfaces over a wide range of height, the recurrent suspicion that they are marine, rather than sub-aerial, and the suggestion of a eustatic interpretation of the evidence, bids fair to make it even more difficult to think in terms of 'cycles' in the orthodox sense.

The foregoing argument is intended to indicate some of the reasons why geologists tend to ignore geomorphology, especially in Britain. It is not implied that the reasons are entirely sound or that no counter case could be put. It may be conceded, however, that the geologist, confronted with the need of specialist

[1] Davis, *op. cit.* (1895) and p. 97 of the present volume

training in many different fields, can only be expected to accord place to geomorphology in so far as it pays dividends in illuminating earth-structure and earth-history. Regarded as a special branch of latter-day growth, it is only one of several ; geophysics and micro-palaeontology suggest themselves as modern developments affording powerful tools in elucidating earth-history, quite apart from their immediate economic applications. It is easy to say that the whole range of possible methods of investigation should be applied to the problems of geology, but foolish to ignore that they are of very different powers and values. Geomorphology is a part only of physical geology and a smaller part still of the whole science. Nor can we expect it to be cultivated solely at the behoof and for the service of geographers.

Our brief review of the relations of geology and geomorphology illustrates a principle of great importance in what I have referred to as the 'sociology of knowledge'. A subject is best understood as a whole, not by defining its logical boundaries, but by considering the methods or tools by which its aim is pursued. It is in the light of this principle that we shall find it profitable to examine the relations of geomorphology and geography.

Historically, these relations are essentially simpler than those we have traced in the case of geology. Geomorphology emerged as something like a coherent subject towards the end of the nineteenth century at the very time that academic geography was in process of being re-born. In terms of Mill's analogy, geomorphology stood to geology in the relation of 'an edifice which had afterwards sprung up'. To the new geography it was one of a number of 'aggregates already in existence', which, it seemed, might conveniently and profitably be circumscribed by the geographical wall. Considerable efforts were made thus to include it by Peschel, Richthofen, and Penck in Germany and by Davis and Tarr in America. No similar situation can be said to have arisen in Britain ; geography here was initially too weak to circumscribe anything, and in any case its British founders had other ideas. One has only to compare three volumes in the Oxford Regions of the World Series, Mackinder's *Britain and the British Seas* (1906), Partsch's *Central Europe* (1903), and Russell's *North America* (1904) to see the varying emphasis on geomorphology fully illustrated. Mackinder had, no doubt, the most compact and manageable region to survey. His book deals

88

adequately with the geomorphology ; indeed it is not too much to say that it embodies original contributions in the field, notably the chapter on the rivers of Britain. Yet the British study accords to geomorphology a relatively subordinate rôle. It was perhaps his experience in seeking unavailingly to teach his 'regional method' to other contributors to his series which led Mackinder, many years later, to aver that geographers 'in escaping from servitude had robbed the Egyptians, the geologists, and had been cursed for the possession of ill-gotten goods by a generation spent in the wilderness'.[1] Yet by the time that it was uttered the reproof had lost much of its force. Geographers were looking increasingly askance at the attractions of geomorphology and were seeking, by ignoring it, to throw it back, as unwanted and irrelevant lumber, into the geologist's back garden.

I shall not attempt to trace fully here the rambling and indecisive controversy on the place of geomorphology in geography. The clearest recent statement is that of Ogilvie in his Herbertson Memorial lecture.[2] He affirms the need of close study by geographers of land-forms and declares in favour of 'explanatory' rather than empirical description. He notes that geologists in Britain present, as a rule, only vague and elementary statements on the origin of relief, and in effect abandon the last chapter of earth-history. He feels it will be necessary for some time yet for geographers to be trained in morphological methods, so that they can make good, for their own purposes, the omissions of geologists, but he urges them to remember that in so doing they will tend to write evolutionary studies, essentially chapters in earth-history, which, in his view, 'ought to come from the pen of the geologist'.

There is much in Ogilvie's treatment which commands assent from all geographers of reasonable mind, but a number of criticisms may fairly be made. The 'genetic' difficulty is, indeed, an old one. An extreme interpretation of the nature of geography would fix its whole attention on being rather than becoming, but this represents an ideal impossible of attainment, and of curiously little intellectual attraction when attained. If to describe land-forms is geography and to discuss their origin is geology, then a

---

[1] *Report Brit. Assoc. Adv. Sci.* (London, 1931. Pres. Address to Sect. E), p. 4   [2] *Geography* 23 (1938), 1

similar limitation must apply throughout the field of systematic geography. Hartshorne enforces his recent discussion of 'The Nature of Geography' [1] by a diagram in which the 'geographical plane' intersects the plane of the systematic sciences, each of which has its geographical counterpart—meteorology—climatology, botany—plant geography, economics—economic geography, and so on. Yet in each case the 'genetic problem' exists. Plant communities are subject to 'succession' and human societies to development. Climates are the result of genesis not less than land-forms. One notes that Hartshorne omits geomorphology from the geographical plane altogether, substituting the simple words 'land-forms'. This no doubt expresses his conviction that geomorphology is really geology, and had, therefore, better not be mentioned. Reflection assures us that this is an impossible and ultimately a disastrous position to take up. To pursue it logically would be to concede that as climatology, economic geography, and the rest become genetically coherent they too will pass out of the field of geography and that it will be impious to mention them! The truth of the matter surely is that it is quite impossible for geography to achieve an *instantaneous* cross-section of the phenomena of the systematic sciences. All the phenomena are subject to 'becoming' and their inter-relation cannot be perceived if their evolutionary tendencies are ignored. The element of truth in the implied objection is that if the geographer moves too far in the dimension which lies, as it were, at right angles to his own, parts of his field will disappear over the curve of the mental horizon and his synthesis will become unbalanced. But this is only to say that he must limit himself to consideration of proximate, not ultimate, genesis. Whether considered as a whole, or in any of its branches, geography must remain, to this extent, a two-dimensional not a one-dimensional subject.

Geomorphology is the historical geography of the physical landscape. The only alternative to the genetic approach is to accept land-forms as 'given,' which would be in effect to see them with dull and unimaginative eyes and in all probability to misread them and misconceive their relations. Davis tersely and finally vindicated the genetic approach when he wrote : 'To look

[1] *Ann. Assoc. Amer. Geog.* 29 (1939), 323

upon a landscape . . . without any recognition of the labour expended in producing it or of the extraordinary adjustments of streams to structures and of waste to weather, is like visiting Rome in the ignorant belief that the Romans of today had no ancestors.' [1] It is the idea of development or evolution in the sense here implied by Davis which manifestly constitutes the major philosophical contribution of geomorphology to the bare bones of physical geography. This is undeniable—yet it has been denied, at least by implication, by certain geographers. It is with some feeling of surprise that I find myself standing here, in the college where Lyell first professed his subject, attempting to reaffirm a principle so self-evident. For geography to reject the intellectual concept of developing land-forms would be stupidly retrograde, the self-imposition of limitations inevitable in the days of Humboldt and Ritter—but quite unnecessary in our own. Geography has, indeed, shown from time to time a tragic propensity to simplify its domestic regimen by 'throwing the baby away with the bath-water'. It has shown this not in regard to geomorphology alone ; some of its reformers propose thus to serve a veritable platoon of babies with their appropriate *quanta* of bath-water. Given their way they would zealously proceed to empty geography of most of its content, value, and significance. I can only conclude that, in the language of modern psychology, they have become possessed, in respect of their subject, by a psychotic death-urge ; indeed, their persecuting zeal is such that, if uncurbed, its life would very rapidly become extinct.

The more immediately practical relevance of geomorphology to the geographer is seen in two distinct, but not unrelated, con-texts. In the first place, though processes of landscape evolution are slow in comparison with those of human development, the physical evolution of what are essentially human regions is still proceeding. The past becomes the key to the present in con-ceiving of these regions as wholes. The great Aquitaine Lowland lying between the Pyrenees and the Massif Central of France, was once a gulf of the late Tertiary sea, almost literally an extension of the present Bay of Biscay ; the gulf was gradually filled up with rock-waste from the surrounding uplands. This process is

[1] *Geographical Essays* (Dover, 1954) p. 268

still actually in progress by virtue of the heavy river deposition of the Garonne and its tributaries ; it affects the life of the present inhabitants of the region, leaving its stamp upon land-use, settlement, and economy in general. The physical history in this and many other cases is more than mere background to the human development ; it overlaps with and is continued in it. While geography as practised by geographers retains its root meaning of a 'description of the Earth' it can no more be blind to physical than to human history.

But again, when we come to subdivide areas in detail, to examine the actual differentiation of the Earth's surface, the first of our considerations is generally its morphological diversity. Physical 'regions' may be distinguished on many different bases, climate, structure, lithology, soil, or hydrology. Where a cover of natural or semi-natural vegetation exists this is, no doubt, the best index to the natural variation of the surface. Where a true cultural landscape has replaced the primeval scene this method fails us ; it fails us, indeed, in varying degree throughout the greater part of the world. Morphology is, then, not only the factor which, taken alone, most readily integrates the others, but the morphology, even the micromorphology, generally shows very close correlation with the warp and woof of the cultural landscape. The student is too apt to assume that when we distinguish, say, the Cotswold terrain, we are dealing essentially with a physical region. It is certainly this, but it is evidently more than this. The detailed texture of the human landscape is a faithful reflex of the form of the surface. So is it likewise, in the familiar and classic French *pays*. Again, we may traverse a Yorkshire dale and find all the road and rail crossings on morainic bridges, the settlements on morains or late glacial deltas, and the intervening lake flats forming natural farm units. The Chalk cuestas of the South Country offer their recurrent pattern of scarp-foot and dip-foot 'sites', using the term 'site' as the pedologists use it, but extending it in the wider geographical context. The dip-slopes vary in form, soil cover, and utilisation according as they have been exhumed from beneath an Eocene or a Pliocene cover, or have been exposed since Mid-Tertiary times. It is in the light of such examples as these that the geographer studies geomorphology, not because, or not only because, it is a register of Earth history, but because it is a guide to land quality in the widest sense. Every physiographic fact is, indeed, the starting-point of two trains of inference, one leading back into the geological past and one into the geographical present. No doubt it is true that a geographer's choice of tools and the very philosophy of his subject are conditioned in some measure by his environment ; certainly no British geographer can readily deny the relevance of geomorphology, But there are in fact very few areas of the Earth's surface in which the form of the land is not the best and most immediate indicator not only of its recent history but of its structure, variations of soil, and vegetation cover, and the mode of circulation of water above or within it. When human occupancy and economy are also found to show significant relationship with the facts of form, surely the geographer may conclude that morphological analysis is fundamental to his final synthesis and value it accordingly.

Again there may come the query—why cannot we accept the facts of form without studying their antecedents ? To this question we have already given, in the word of Davis, the general philosophical reply. But it prompts other reflections. I have heard it innocently and plaintively suggested that if only geologists would really bend their minds to the description in detail of the geomorphology of the world, regional geography would be much easier to study. It is not long since Mrs Barbara Wootton [1] made effective play, in another connection, with the nursery poet's lines :

> If all the world were apple pie,
> And all the seas were ink,
> And all the trees were bread and cheese,
> What should we do for drink ?

The format of the argument fits our case. If all geologists were to turn into geomorphologists, all botanists into ecologists, if all historians were knowledgeable cartographers and all economists were preoccupied with the geographical implications of applied economics, geographers might well be in search of an occupation ; they could adopt a pose of delightful ease, but one that would soon cease to retain any dignity or credit whatsoever. But none of these admirable labour-saving devices is at all likely to avail the geographer. He cannot sit back and receive his data like some pre-digested intellectual breakfast food. He must

[1] In *Lament for Economics* (Allen & Unwin, 1933)

needs go out and fetch it himself and adapt it to his purposes. In order to do so he must at least be trained in the necessary ancillary disciplines. It is at this point that so much confusion seems to be engendered. However regrettable it may seem to some when the research geographer is lured into evolutionary studies, one cannot artificially break the chain of relationships in training and return a pedantic methodological negative if the student asks 'why?' Indeed, my humanist colleagues among the geographers never dream of such a negative within the social field. They blandly trespass at will, as fancy directs, and I for one will not say them nay if they will be logical enough and tolerant enough to extend a comparable freedom to my co-workers and myself.

The argument, as I have sought to present it, may be summed up by saying that as a contribution to Earth history the evidence of land-forms is vastly more incomplete and fragmentary than that of stratified deposits, but as constituents of landscape, land-forms manifestly take a primary place. There are innumerable areas in which the record of land-sculpture can add little to geological history, but which for the geographer are important regions or parts of regions. In describing the land-forms of such an area he will necessarily indicate how they have developed and are developing, and in so doing he need not be deterred by the pedantic objection that he has for the nonce become a geologist. If in the same area he traces changes recently occurring, and perhaps still continuing, in the distribution of cultivation or settlement, no-one in his senses will accuse him of becoming a historian !

It cannot surely be an unreasonable or improper hope that geologists may not turn their backs on geomorphology as too 'geographical' nor geographers shun it as 'geological', but that both may cultivate the field for their proper purposes. If in so doing they on occasion help one another this will be a matter neither for surprise nor regret. Sovereign subjects are perhaps as outmoded in the world of today as sovereign states, and above the jealousies and disputations of the separate sections of the learned world stands the truth that knowledge is one and indivisible.

7

The Changing Physical Landscape of Britain

IT IS more than half a century since William Morris Davis, having learned the physiographic lessons of the American Southwest and applied them in the Appalachian country, visited Britain. His visit revived interest in British land-forms, at the same time giving it a new direction. In his famous paper published in the *Geographical Journal* in 1895 he sketched the physiographic evolution of southeast England.[1] The central feature of his interpretation was his recognition of the widespread hill-top plain of the area regarded by him as an example of an uplifted and dissected peneplain, his own term for that undulating surface which marked the termination of a long-continued cycle of erosion.

Let us recall in imagination that most famous and extensive view of southern England which we obtain from the tower on the summit of Leith Hill. Here, just over 1,000 feet above sea-level, our prospect embraces a dozen counties. The natural horizon line is some forty miles distant, but the actual view, given maximum visibility, falls short or ranges farther than this with the accidents of relief. What renders the view signal and notable is the fact that the major hill ranges of the south country lie, for the most part, near the horizon circle and thus frame the regional picture. To the northwest the skyline is the level crest of the Oxfordshire Chilterns, continued northeastwards to the Dunstable Downs and, on a clear day, even beyond the Hitchin Gap to the Chalk ridge south of Royston. Westwards and southwestwards the view is restricted by the high bastions of the Lower Greensand at Hindhead and Blackdown, but behind them the western Downs above Selborne present a crest as level and featureless as that of the Chilterns. It is continued in the summit line of the South Downs and the high central Wealden crests of Ashdown Forest. In the northern foreground, seen in much greater detail, the North Downs present a similar even summit.

[1] 'The development of certain English rivers' *Geog. Journ.* 5 (1895), 127–46

In these facts lies the evidence for the theory that the south country has been sculptured from an uplifted plain of which relics are preserved in our highest hill summits. A plane 800 feet above sea-level would, indeed, pass close to all these summits, Leith Hill (965 ft.), Blackdown (918 ft.), and Hindhead (890 ft.) would exceed it, as would also the highest points of the North Downs (877 ft.), the South Downs (880 ft.), and the Chilterns (852 ft.). Crowborough Beacon (792 ft.) just fails to reach it,

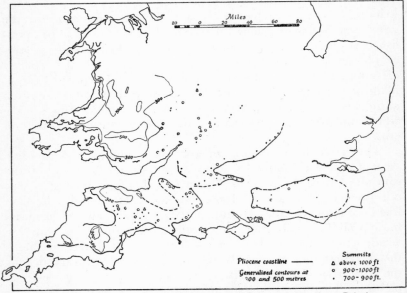

Fig. 1   Pliocene coastlines and ancient summit levels in southern Britain

as do many parts of the Chalk crest-lines. These summits, and others over a wider area, are shown in Fig. 1.

This peneplain was regarded by Davis as the product of the age-long waste and river action which ensued in mid-Tertiary times, after the strong Earth ripples, distant ground swell of the Alpine storm, had thrown the south country from Kent to Somerset into parallel ridges and furrows and thus given birth to the first decipherable drainage system in the area. These first rivers produced the plain ; thereafter, as it was uplifted, they dissected it, bringing into being the graceful lines of the present landscape with its broad lowland vales separated by scarp-bounded uplands through which the rivers, older than the hills they traverse, wind in narrow defiles.

Davis thus recognised an ancient land surface. In its later development, physiographic work in the southeast has become a study of this and other surfaces of differing dates and various degrees of preservation. There is indeed a true geology and geography of surfaces, distinct from, though not unrelated to, that of former sea-bottoms and their wide sheets of stratified sediments. The geomorphologist in seeking to take on the tale from the hands of the stratigraphical geologist finds himself, in the words of T. H. Huxley, seeking to use 'sea reckoning for land time' and the difficulties of so doing are formidable. The hill-top plain of southern England may be reckoned, in this sense, a Miocene surface; the product of atmospheric action during that Miocene period which, alone among the major time divisions of the geological past, is unrepresented by stratified rocks in Britain.

If we are to use the southeast as a clue and a measure for other British regions we must take note of the fact that two other well-marked surfaces can be traced in its landscapes. Younger than the main hill-top plain is the broad bench, cut largely in Chalk, on which the remnants of marine Pliocene deposits are widely distributed. This was not treated by Davis, for, though Pliocene submergence was in his day known to have occurred in Kent, the full extent of the sea was not then recognised. The character of the deposits themselves, their occasional fossils, and above all the evidently wave-cut flat on which they rest and which is recognisably preserved even when the deposits have been removed, have since enabled us to trace the former extent of the sea. In making the reconstruction shown in the map, Professor Linton and I were guided by a further important consideration—the relation, or rather lack of relation, of much of the drainage to the structure of the underlying rocks.[1] When England south of the Thames was first raised into closely spaced ridges, the main lines of drainage must have followed the furrows between them. But that is not at all what we find over much of the main belt of folding today. Frequently and systematically, north–south rivers cross the east–west fold axes at their strongest development.

---

[1] *Structure, Surface and Drainage in South-East England* (Philips, 1955), pp. 48ff.

This is notably true in Sussex and throughout the Wessex Chalk area, where it was first clearly perceived by Linton. It was a major defect of the Davisian hypothesis that it failed to account for such features. It is true that the first cycle of erosion would have obliterated the anticlinal crests, but if the present drainage were merely successor to that of the first cycle it could not, save by the local accident of river capture, have so developed as to cross the line of strong anticlinal folds. What is required to account for the facts is a new start for the drainage such as must have followed the spreading and subsequent uplift of a sheet of sand and shingle over parts of the old plain. It is worthy of note that it was Davis's student and chief successor, Douglas Johnson, who first hinted to us that an extension of the Pliocene sea into Wessex would satisfactorily account for drainage features otherwise hardly explicable.

Throughout the greater part of the area the Pliocene shelf or platform is backed by a low bluff separating it from higher ground—part of the mid-Tertiary plain. This bluff fixes the position of the coastline.[1] It is a feature of the greatest significance that the platform maintains a very uniform general level of about 600 feet. It slopes seaward from between 650 and 700 feet at the base of the bluff to about 550 feet or a little lower. This shows beyond question that the late Tertiary uplift of much of southeast England was uniform, i.e. unaccompanied by faulting or bending. On the Essex coast, however, and northwards into East Anglia, Pliocene deposits contemporary with and later than those of the London area lie but little above sea-level ; in Holland they are far below it. Thus is exemplified one of the later expressions of a long-established and recurrent tendency for the east coast region to be tilted towards the North Sea depression.

Excluding this eastern zone, the physiographic development of the area has consisted in the uplift and dissection, stage by stage, of the Pliocene sea floor and the adjacent parts of the mid-Tertiary plain. The stages of this uplift are marked by the minor

---

[1] It is particularly to be noted that in the central Wealden area, which is regarded as lying beyond the limits of the Pliocene trespass, the drainage retains its dominantly longitudinal lines. This provides confirmatory evidence and guides the reconstruction, but since there is no coastal bluff here the argument must not be permitted to beg the question.

platforms and terraces, essentially old valley floors, of which the district presents a complex but coherent record. The chief of these are marked by surfaces at about 400 feet north of London [1] and, more widely spread, the dissected plain, now about 200 feet above sea-level. This was formed before the main ice advance into the region, for much of the glacial and associated deposits rest on it. [2] Dissection has, however, revealed a regionally much more important surface, the early Tertiary plain on which the main mass of the Tertiary (Eocene) deposits rest. The Chalk sea floor was raised into land and subjected to long erosion before the earliest Tertiary deposits were laid down ; it must have been far advanced towards peneplanation before the waves of the Tertiary seas completed the levelling process. It is sometimes exposed in deep excavations through the Tertiary cover, but it has been laid bare by the clipping back of this cover, on the margins of the London and Hampshire basins. It necessarily shared in the Alpine movements and thus is, normally, tilted at a considerable angle.

If we turn now from southeast England towards the uplands of the west and north there are thus three main surfaces, the early and mid-Tertiary plains and the Pliocene marine bench, which we might hope to find there represented.

Let us first pass westwards beyond Abbotsbury where the restored Pliocene coastline crosses the present coast. In east Devon a similar feature recurs as part of a whole series of platforms. If we traverse southwards from Honiton along the ridge of Farway Common and Broad Down and thence to the coast near Branscombe, we descend a gigantic physiographic stairway each of the treads of which was interpreted by the late J. F. N. Green as a marine platform. [3] At the highest level, at about 800 feet, is the great plateau which also extends ten miles north of Honiton to the scarp of the Blackdown Hills. This surface might well commend itself as the continuation of the hill-top plain of the southeast, cut here in the Upper Greensand. Green, however, regarded it as an old sea floor, finding traces of rolled shingle upon it at Ottersford and elsewhere. South of Broad Down we descend to a lower surface at about 650 feet and roughly a mile

---

[1] S. W. Wooldridge, *Proc. Geol. Assoc.* **37** (1927), 107
[2] *Ibid.* **39** (1928), 1   [3] *Proc. Geol. Assoc.* **52** (1941), 36

wide. This is replaced as we approach the main Exeter-Lyme road by still another with its back at about 600 feet. Nearer the coast a narrower flat comes in at about 530 feet, ranging almost to the cliff top.

If we group together the two surfaces between 700 and 530 feet, they form a feature very similar to the Pliocene bench farther east and the coastline of the time can be tentatively reconstructed as following the main bluff below 700 feet. It is, however, clear that we are offered an almost embarrassing number of such surfaces, which if they are correctly interpreted as marine strand-flats plainly record a shore line retreating during mid-Tertiary times under the influence of long-continued but spasmodic uplift. We select this example for special mention because here, in east Devon, we are still within the province of the rocks of the English lowland. All these surfaces are cut in Upper Greensand and are necessarily at least post-Cretaceous in age. Comparable regular sequences of platforms have been shown to occur below 1000 feet throughout much of our southwestern and western coastlands, in the Bristol district, Devon and Cornwall, South and western Wales, the south and west coasts of Ireland, and the Lake District. They present a difficult and very disputable body of evidence. I have recently reviewed and discussed this evidence as a whole and will not enter fully into it here.[1] The most effective brief summary of the facts has been given by Hollingworth [2] who made a statistical analysis of the hill-summit heights of large parts of our western uplands. He found frequency maxima at the following chief heights : 320 ft., 430 ft., 560 ft., 770 ft., 900 ft., 1040 ft., and 2000 ft. His method cannot distinguish between marine and river-cut flats, but the results certainly suggest the conclusion that much of Britain has suffered spasmodic uniform uplift. This is of course to make the large assumption that coastal flats at the same altitude but at widely separated points are of the same age. Our present evidence may suggest this but it certainly cannot prove it.

We may pause in our westward traverse on the summit of the Haldon Hills, south of Exeter. Here we are still on the Upper Greensand and our summit is evidently a continuation

---

[1] *The Advancement of Science* 7 (1950), 162 ; reprinted as Essay No. 9 of this volume.　　　[2] *Quart. Journ. Geol. Soc.* 114 (1938), 55

of the 800-foot flat of east Devon, but we are less than twenty miles from the high summits of Dartmoor which present their characteristic profile from across the deep intervening rift valley of Bovey Tracey. Devon and Cornwall continue and supplement the physiographic record of our lowlands, but they introduce also the very different problems of our uplands. Unglaciated, they have borne the full brunt of Tertiary uplift and erosion. Here, if anywhere, we may hope to find the essential clues to the history of the British land surface, but the record is hard to read and much work remains to be done. In 1909, using Devon and Cornwall as a comparator in his study of glacial erosion in North Wales, W. M. Davis interpreted the landscape in terms similar to those he had used for southeast England.[1] The granitic uplands stood as residuals above an uplifted peneplain, the product of an earlier cycle of erosion but now dissected, with lowlands opened up along the softer rocks of the eastern margin. Only a year before, George Barrow had proposed a very different alternative line of interpretation in his paper on the high-level platforms of Bodmin Moor.[2] Here and on the flanks of Dartmoor we find extensive flats cut indifferently across the structure and maintaining for long distances altitudes of about 1000 and 800 feet respectively. The 400-foot coastal flat, prominent throughout Cornwall, was already well known and linked, though rather dubiously linked, with fossiliferous Pliocene deposits. Its precise age is still disputable, but it is beyond question a marine strand-flat backed by a degraded cliff. Barrow interpreted the higher platforms as similar but older features. We note at once the absence of a 600-foot platform from the sequence ; its presence has, however, been claimed by Balchin on the flanks of Exmoor and by Green in the region of the Dart valley. There are also clear signs of its presence along the northern side of Dartmoor and the southern side of the Bodmin upland.

In terms of platform sequence the terrain of older rocks in Devon and Cornwall can thus sustain a similar interpretation to that of the east Devon plateau region. When we turn, however, to greater altitudes on Dartmoor and Exmoor new problems emerge. Is it here that we may hope to find relics of one or

[1] *Quart. Journ. Geol. Soc.* **65** (1909), 291
[2] *Quart. Journ. Geol. Soc.* **64** (1908), 384

other of the two great Tertiary plains of the southeast? Before making such a claim we must recall that the lowland rocks, notably the New Red Sandstone and the Chalk, must once have extended widely beyond their present limits in the southwest. We are left thus with the question, how far the arid climates of New Red Sandstone times or the waves of the Upper Cretaceous sea may have left their mark on the surviving high country of the southwest. To this question we shall attempt, for the moment, no direct answer. Let us adopt rather the strategy of indirect approach, returning to southeast England and passing thence *via* the Cotswolds to Wales.

Standing at 856 feet at Uffington Castle on the Berkshire Downs, our view ranges across the upper Thames valley to the Cotswold summits forty miles distant. Here at Cleeve Cloud and Broadway Hill the scarp crest just exceeds 1000 feet. These points, taken with the other high Cotswold summits, seem to indicate that we have again found remnants of the great mid-Tertiary plain. We must note, indeed, that the gap between the Chalk and Oolite crest is a wide one and unbridged by linking summits. If, however, we entertain the not unreasonable hypothesis that the two crests are, broadly, relics of the same surface we naturally turn northwestwards where our view embraces such summits as Worcestershire Beacon (1395 ft.), Brown Clee (1790 ft.) and the Stiper Stones (1731 ft.). If we seek to include these heights in our former plain we find that its rate of westward climb is accelerating ; but taken by themselves we must admit that these summits are too few and far between to subserve much more than idle speculation. From the Stiper Stones, however, we may get a magnificent prospect of that much more impressive feature, the high plateau of Wales. In North Wales its general level is about 2000 feet. Above it Plynlimon (2468 ft.), the ring of peaks around the Harlech dome (Cader Idris, the Arans, and Arenig) and those of Snowdonia rise distinctly. South of Plynlimon the general surface declines gently southwards, reaching about 1500 feet in Mynydd Eppynt where it is overlooked by the scarp of the Old Red Sandstone, in its culminating points overtopping 2000 feet for a distance of nearly forty miles. Southwestwards the descent is even longer continued ; there is no obvious break until a level of 1000 to 1200 feet is reached in north Pembrokeshire where the Prescelly Hills (1760 ft.) rise suddenly from a generally lower surface, just as do the high peaks of North Wales.

There has been wide divergence of opinion as to the age and origin of the Welsh plateau ; here indeed the issue is fairly joined between two radically contrasted modes of interpretation. Ramsay first recognised the great surface a century ago and saw it as a plain of marine denudation. If marine denudation is in question we are faced with our ignorance of the extent to which marine Jurassic formations may formerly have covered Wales. A more likely agent of planation would however be the waves of the Chalk sea ; this was the view taken by A. J. Jukes-Browne [1] who, in effect, regarded the higher standing residuals as relics of islands in the Cretaceous sea. But in 1909 Davis [2] proposed an early or mid-Tertiary date for the surface, inclining to the latter alternative. He was making the correlation which our westward view from Broadway Hill might suggest, and claiming once more that the hill-top plain was a mid-Tertiary feature as in southeast England and in Devon and Cornwall. In the following year W. G. Fearnsides regarded the surface as probably an early Tertiary surface of sub-aerial denudation. Twenty years later in a most important address O. T. Jones,[3] while admitting the possibility of Jurassic or Cretaceous marine action, stressed the great efficacy of planation under the desert climates of New Red Sandstone times. He quoted Davis's paper on the cycle of arid erosion and the work of Passarge in South Africa which showed that the desert plains are 'not undulating, with low hills, but true plains of great extent from which isolated residual mountains arise like islands from the sea'. To such a description the Welsh plateau certainly conforms satisfactorily though it bears evidence of subsequent tilting. Professor Jones has recently returned to the subject in an address on the drainage system of Wales.[4] He still regards the high plateau surface as primarily due to the New Red Sandstone erosion but accepts the hypothesis that the drainage began on a Chalk cover, and thinks it 'not improbable that the uplift of the area . . . may have begun in the early

[1] *The Building of the British Isles* (see p. 11, Standford, 4th ed. 1922), p. 334
[2] Davis, *op. cit.* (1909)
[3] 'Some episodes in the geological history of the Bristol Channel region' *Report Brit. Assoc. Adv. Sci.* (Bristol, 1930 [1931]. Pres. Address to Sect. C)
[4] *Abs. Proc. Geol. Soc. Lond.* No. 1475 (1951)

Tertiary but attained its culmination in the Miocene'. He also finds evidence of a lower surface along the main drainage lines, controlled by a base level at about 600 feet, and a coastal surface similarly related to a 400-foot base level.

The latest paper that bears on our problem is that of D. L. Linton,[1] which is concerned to define the form of the original land block from which Scotland has been carved, In the Scottish Highlands, Ben Nevis and the great Cairngorm plateau rise to over 4000 feet, above the 'Grampian main surface' at about 3000 feet. Linton adopts the working hypothesis that these summits mark fairly closely the level of the base of the former Chalk cover. Pursuing this line of argument southwards, he has distinguished each summit 'which is the most easterly of its altitude in its own latitude and is not flanked by higher summits north and south on its own meridian'. Between the plotted points rough form lines have been drawn, and these define a generally easterly slope from Caithness to Devonshire. Such a slope is consonant with the fact that the original direction of the drainage over most of Britain appears to have been easterly.

We cannot discuss here the many implications of Linton's stimulating paper. It must suffice to note for our purposes two aspects only. In Wales, Linton's reconstruction is consistent with either of the leading views on the Welsh high plateau. His form lines are drawn through the residual summits, and if these mark the base of the Chalk they leave room in most areas for a sub-Cretaceous cover of Mesozoic rocks resting on a Palaeozoic floor produced by arid or marine erosion. If such were the case we should conclude that the vast periods of Tertiary erosion spent themselves in stripping this floor of its cover and exhumed it largely intact. If, alternatively, following Davis, we infer that Tertiary planation itself has left its mark on the present face of Wales, removing a Chalk cover and cutting a plain below its base, such inference is not inconsistent with Linton's reconstruction.

Farther south difficulties arise ; Linton includes in his control points the summit of Cleeve Cloud on the Cotswolds. It is unlikely, though perhaps not impossible, that the Chalk ever rested here ; but in any event one cannot safely assume that its

[1] *Scot. Geog. Mag.* **67** (1951), 65

own lower member, the Gault-Upper Greensand, was missing. This, however, hardly involves more than a minor correction. It is more important to notice the large divergence of view implicit in regarding this point alternatively as on a sub-Cretaceous and a mid-Tertiary surface. Still farther south the lines of Linton's reconstruction cross Dartmoor. The Dartmoor summits certainly conform to a plane sloping southeastwards at rather less than half a degree, but J. F. N. Green [1] has given reasons which appear sound for regarding this surface as of early Tertiary date, and there are other strong reasons for doubting whether the Cretaceous base can here have had this simple form. It was necessarily involved in the strong Alpine folding, certainly traceable as far west as Somerset, not to mention the considerable faulting which gave rise to the Bovey Tracey rift.

Space will not avail us to discuss here the problems of the Pennine upland. It must be sufficient to say that they are in all respects similar to those of Wales. On the one hand the high summits have been claimed as representing one or more Tertiary peneplains, while in the south the general upland plain of the Peak District has been supposed to represent the sub-Triassic surface.

The objects of this paper have been definite and limited. It seeks to indicate though not to solve some of the leading problems confronting British geomorphologists and to illustrate the methods and arguments employed by them. Professor Darby, speaking as an historical geographer, recently sketched in broad review the changing British landscape in its human aspects.[2] Geomorphology is best seen as the historical geography of the physical landscape and, despite great differences in material and methods, the two studies show a strong basic kinship since both deal in terms of sequence and require an evolutionary approach. The chief difference between the two studies is, however, their own present state of evolution. Work on the changing human landscape in Britain is still in its earliest stages—it is a project clearly in view and actively in progress. The similar project for the physical landscape was announced and initiated by Davis fifty

[1] *Proc. Geol. Assoc.* **60** (1949), 105 *et seq.*
[2] *Geog. Journ.* **117** (1951), 377–98

years ago. His lucid pen and graphic pencil bred a lineage of workers now in its second generation. A. J. Herbertson speaking to O. J. R. Howarth in 1901 referred to 'this whole great new science' of geomorphology.[1] The claim was scarcely too high, and that it could be made at all was due essentially to the work of Davis. His American successors have recently reviewed and appraised his work.[2] If I venture here to add this brief tribute from Britain it is because of the great debt that many British workers owe to him.

The record of the changing physical landscape of Britain proves, indeed, to be more complex and disputable than is suggested by the writings of Davis. The earlier British emphasis, not unnatural or inappropriate in our sea-girt island, was upon marine denudation. In part this proved demonstrably in error. Our lowland escarpments are not old cliff lines, though it is worthy of note that we have not yet reached the centenary even of this conclusion. That late Tertiary marine planation has left legible marks on our coastlands below 1,000 feet is still a reasonable hypothesis with much evidence to support it.

The other chief point at issue has also been indicated in this paper. The lowland summits are parts of a surface which Davis and others have regarded as a peneplain of mid-Tertiary age. When we seek the corresponding surface in our uplands we confront the possibility that their own even skylines preserve evidence of much earlier phases of planation, notably that of New Red Sandstone times. This suggestion finds support in our growing realisation of the vigour and facility of plain formation under the control of arid or semi-arid climates—a fact widely demonstrated in Africa. If our uplands retain relics of surfaces so ancient, it follows that they must register all later earth movements, so that in general they should be tilted or warped. Some of them indeed show clear evidence of such warping. But in some cases at least we may be justified in supposing that the surface marks the early Tertiary plain (p. 99) and that the warping was part of the Alpine movements. In general, the case for early Mesozoic planation in our uplands must be considered strong but unproven. It requires the unlikely but not

[1] *The Advancement of Science* 8 (1951), 158
[2] 'Symposium in geomorphology in honor of the 100th anniversary of the birth of William Morris Davis' *Ann. Assoc. Amer. Geog.* 40 (1950)

impossible assumption that the great periods of early and middle Tertiary erosion peeled the later sedimentary skin from the uplands but left their core virtually untouched. It is perhaps not unforgivable to regard this proposition with some reserve. If we accept it, and consent also to the claim of those who see evidence of marine planation as high as 1000 feet in the western coastal areas, there is no room left for the Davisian peneplain ; if it ever existed it has been completely destroyed. As J. F. N. Green wrote in his last paper, 'it would seem that the peneplain must ultimately disappear altogether from the English landscape'.[1] Yet this may well prove a premature conclusion. We can return to our first view-point on Leith Hill and see a surface which is certainly Tertiary in age and yields no evidence whatsoever of marine origin. The essentials of W. M. Davis's view may yet be vindicated.

It will be clear that we have necessarily left untouched many cardinal aspects of our changing physical landscapes, notably the developing drainage pattern and the contribution of glaciation. It is true indeed that we have been preoccupied with the earlier stages of our landscape evolution. But it is as true of physical as of human history 'that nothing less than the whole of the past is necessary to explain the whole of the present'. As James Hutton long ago wrote, on the sequence of terrestrial change, 'in examining things which actually exist and which have proceeded in a certain order it is natural to look for that which has been first'. The primary questions we have posed require an answer before we can see our changing physical landscape in its correct proportion and its full sequence.

[1] Green, *op. cit.* (1949)

8

The Physique of the Southwest

IN MANY signal respects the southwestern peninsula stands out among the physical regions of Britain as highly distinctive, if not unique, in its form and evolution. Here is a country of 'upland plains', of level sky-lines and slopes gentle and regular, save along the margins of the deeply incised river-trenches. Yet, as is well known, the land-forms are the long-term successors of a major mountain system. Late Palaeozoic times saw the crumpling and over-thrusting of some 10,000 to 15,000 feet of Devonian and Carboniferous sediments and the building of an important section of the great Hercynian or Armorican mountain system. The mountain structures are still laid bare in cliff section and valley-side quarries, but the original mountainous relief has long since been obliterated by erosion. The essential 'grain' of the country still shows itself locally in low ridges, or in valleys following fault-lines or other zones of weakness, but the older stages of the physical history of the area throw very little light on its present form. The axis of the peninsula as a whole is oblique to the structural grain, while in detail it plainly presents the results of a long and complex history of river-dissection, punctuated by phases of planation. It is these later chapters of the physical history which fall to the geomorphologist to elucidate, and the aim of the present paper is to give an account of the methods and arguments used by workers in this field and an indication of the conclusions to which such work points.

The Drainage Pattern

Let us make a direct approach to the problems at issue by studying the drainage system in relation to the form of the peninsula as a whole (Figs. 2 and 3).

The main water-parting begins on the cliff-top near Land's End and climbs thence *via* Carn Brea to the chain of high granite summits which lie little more than a mile from the northern coast. It traverses four summits at 700 feet, four at 750 feet, and one at 800 feet before climbing again to 800 feet at Trendrine Hill near St Ives. Here the line bends sharply southeastwards, descending from the granite and crossing the depression which spans the peninsula between Hayle and Marazion. Its lowest point (about 90 ft. o.d.) is at the col south of St Erth. Beyond it traverses Perran and Kenneggy Downs at 250–300 feet *en route*

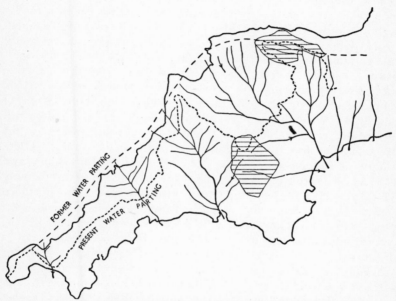

Fig. 2   Reconstruction of original drainage system of southwest England

to Tregonning Hill. This is the only stretch of the line throughout the peninsula which lies close to the southern coast.

From the southern end of Tregonning Hill the line pursues a fairly straight course for some thirty-five miles to near Bodmin, remaining parallel with the axis of the peninsula and not far from its median line, though nearer the northern coast. It first crosses the broad flat-floored col at Nancegollan (300 ft.) and then climbs to traverse the northern side of the Carnmenellis Granite mass, *via* Hangman's Barrow to Carn Marth; the summits in this stretch range between 700 and 750 feet. Beyond, where the line closely follows that of the main Bodmin road, the summit elevations are part of a lower plateau level at 350–500 feet. It rises suddenly to 695 feet east of St Enoder where it passes on to the St Austell Granite, but it quickly leaves this to traverse the outlying twin-summits of Castle an Dinas and Belowda Beacon, both at 700 feet. East of the latter the higher culminating points of the water parting are widely spaced and range in height from 550–650 feet. South of Bodmin the line is notched by a col at 400 feet before beginning its northward climb to the high ground of Bodmin Moor. It ascends over Racecourse Downs following the line of the Launceston road and traversing two summits at 700 feet, two at 800 feet, and one at 850 feet, before reaching 1,000 feet at Hawks Tor. Northwards five successive summits are at 950 or 1,000 feet, but there is a further sudden rise on to 'High Bodmin' at Tolborough Tor (1,100 ft.) and Brown Willy (1,350 ft.). North of the latter the line descends on to the remarkable flat of Davidstow Moor, where it just exceeds 1,000 feet in elevation three times in five miles. There follows a sudden descent to 800 feet west of Otterham, and 850 feet on Tresparrett Moor. Here for the first time since leaving Trendrine Hill the water parting returns close to the northern coast. It then falls fairly rapidly to 580 feet near Jacobstow ; from here to Kilkhampton it is parallel with and close to the coast and manifestly traverses a low plateau surface. In a distance of some eleven miles there are eighteen low summits of which four are at 550 feet, ten at 500 feet, one at 450 feet, and three at 400 feet. North of Kilkhampton a slow climb begins and near Woolley at the head of the Tamar valley the line reaches 700 feet before turning sharply southeastwards, transverse to the peninsular axis between the Tamar basin and the headstream of the Torridge. For a long distance the prevalent summit heights are at 600–650 feet, but south of Halwill junction, the line climbs suddenly to 800 feet near Hender Barrow and the higher points are at 800–900 feet for a further ten miles till a further sharp rise to 1,200 feet takes place on the margin of the Dartmoor granite. Here the line describes a short loop over High Dartmoor *via* Kitty Tor, Cranmere Pool, Hangingstone Hill, and Cawsand Beacon, separating the Okement and Taw drainage from that of the Tavy and the Dart.

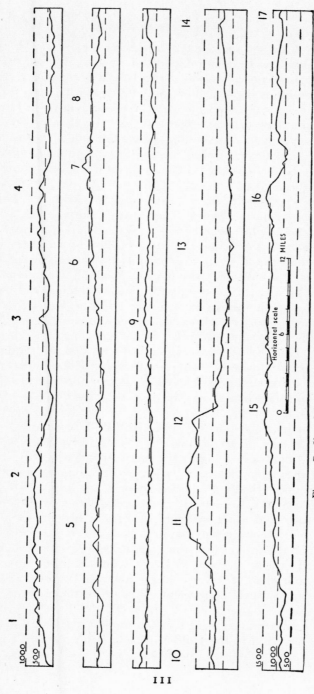

Fig. 3 Profiles of the present peninsular divide in southwest England

III

East of Cawsand Beacon the line leaves the granite, dropping very steeply in less than two miles from 1,750 to 900 feet. Beyond is a slower descent to 550 feet ; two cols at 400 feet are crossed on the Permian rocks near Bow, and after rising again to 550 feet the line descends below 350 feet at Copplestone in the col linking the Yeo and Taw valleys. Northwards a height of 550–600 feet is quickly resumed and a long slow climb then begins up a singularly smooth and unbroken slope to a summit at 950 feet, west of Stoodleigh Beacon. The crest retains levels of 800–900 feet for a further four miles before dropping to 700 feet at Anstey Station in crossing the long furrow which marks the southern edge of the Exmoor upland. Northwards the 900-foot level is quickly regained, but on Exmoor the water-parting loops sharply westwards so as to enclose the eastward-flowing headstreams of the Barle and the Exe. It rises steadily from 1,000 feet north of Anstey *via* the ridge of Anstey and Molland Commons, and turns northwards thence close to the county boundary to Wood Barrow (1,550 ft.) above Pinkworthy Pond at the head of the Barle, where it turns back eastwards *via* the western end of the Dunkery Beacon ridge (1,650 ft.) to the crest line of the Brendon Hills at about 1,300 feet. Here it turns sharply southwards between the southward-flowing headstreams of the Batherm and the Tone, maintaining crest levels at about 1,100 feet southwards to Heydon Hill where a further drop to a level of 900–950 feet occurs.

The main water-parting we have traced is some 250 miles long, and if it could be followed by a 'Cornubian Way' it would offer the geographer a highly instructive journey. It is, in fact, followed throughout most of its length by roads, some modern, but many undoubtedly pre-historic in origin. Despite its variable elevation it marks, in each successive section, the 'height of land' and barrows and other evidence of ancient occupancy abound along it. For our present inquiry, however, its prime interest is the light it throws upon the contest between northward and southward flowing drainage. Every col on the line offers us a possible instance of river piracy and the briefest scrutiny indicates many such adjustments. Three of these are of major importance.

In the far west it is clear that an original southward drainage has been reversed in the Hale-Marazion depression. Today the direct headstream of the river Hayle rises at Germoe, little more than a mile from the southern coast, but the main headstream flowing southwest from near Crowan, and the tributaries which join the Hayle south of St Erth, have directions accordant with a former southerly flow through the depression.

The peculiar looped course of the Camel indicates reversal on even larger scale. The Camel headstream and its neighbour the Allen flow southwestwards from a secondary water-parting close to the northern coast. The trunk stream above Wadebridge makes an angle of about 280° with their direction, and its relation to them and to its main left-bank tributaries suggest that the original drainage was southwards. At Red Moor a broad gap, with floor at about 450 feet indicates the probable course of the old drainage line to the neighbourhood of Lostwithiel and the present Fowey estuary.

The circumstances of this reversal of drainage are suggestively indicated by those of the evidently imminent capture of the headstream of the Tamar by the Bude Haven drainage. The main stream of the Bude embayment trends NNW. Its southward-flowing right-hand tributary, the river Strat, is related to it almost exactly as the Allen is related to the lower Camel. But the steep descent of the streams draining to the coast here must inevitably lead to an eastward migration of the main divide and the beheading of the Tamar ; the most likely place for the diversion would seem to be the col at 350 feet near Whitstone and Bridgerule Station.

This clear case of impending capture which, when achieved, will closely imitate the great Camel loop, throws a similar light on the remarkable course of the Torridge. From its source near Hartland for twelve or more miles the river follows a south-easterly course essentially parallel with that of the Tamar. The high ground of Dartmoor lies athwart the continuation of the line, and it is evidently probable that the original flow was eastwards to the Exe drainage. The lower Torridge from near Hatherleigh to Bideford Bay is either of altogether more recent origin or the reversed descendant of a parallel stream formerly flowing south-east. The line of the Taw continued in that of the Yeo may similarly mark the line of a former stream flowing southeastwards to join an ancestor of the Creedy ; the conspicuous upstream junction with the Taw of its tributary the Bray

flowing southwards from Exmoor evidently supports this interpretation.[1]

The Exe, like the Tamar, still heads close to the northern
coast, its direct headstream the Quarme, continued by the Exe
below Exton, taking a singularly straight line a few points east
of south, followed with local deviations to the estuary. Eastwards
of the Exe, a series of captures has broken the continuity of
similar southward-flowing streams. The Batherm has beheaded
the Loman and the upper Tone has been diverted eastwards
by a subsequent tributary of the Parrett, leaving the Spratford
stream and the lower Calm marking the beheaded trunk and the
Burlescombe gap as a permanent legacy in the route or transport
geography of the Southwest. This capture is analogous to that
of the Bray by the Taw. It has originated by simple adjustment
to structure, the ready excavation of the Triassic rocks. Its western
analogue may well have originated in the same way before the
removal of the Permo-Triassic cover of northwest Devon.

Taking a general view of the rivers as a whole there is thus
clear evidence of a former drainage to the south or southeast, of
which the Exe and the Tamar remain as substantial witnesses,
though other members of the same family have suffered reversal
of their headstreams. If we take account of the reversals it is
clear that at an early date the peninsular water-parting lay
everywhere near, or perhaps even beyond, the present north
coast (Fig. 2). We can carry it in imagination from the northeastern end of the Land's End granite upland across or off-shore
of the Camel estuary to the head of the Tamar, beyond which it
must have crossed the present site of Bideford Bay to follow a line
close to the present northern coast. The drainage system so
reconstructed implies an 'initial surface' sloping southwards.
But there is another element in the drainage pattern which
implies an original easterly slope. This is reflected in the rivers
of Exmoor and in the main stream of the Dart above Holne
Bridge, continued in the line of its former valley east of Ashburton, now abandoned, but directed towards the Teign estuary.
The upper Teign is also an eastward-flowing stream ; it turns

[1] This reversal of the drainage of northwest Devon was perceived long
ago by A. W. Clayden in his *History of Devonshire Scenery* (1906), though his
general picture of the evolution of the drainage cannot now be accepted.

sharply southward near Dunford, but the broad flat-floored col, just below 650 feet near Longcross, suggests that it formerly continued its eastward flow to join the Exe.

The Erosion Surfaces

Our attempt to reconstruct the broad original lines of the drainage system confronts us immediately with the problem of the age and origin of the widely extended, smooth-surfaced plateaux of the area. In brief, the question is whether these surfaces were produced by the rivers now entrenched below them, or whether they were, wholly or in part, inherited ready-made by the removal of former covering strata, or by the uplift of one or more plains of marine denudation.

It is evident that the Permo-Triassic series, or 'New Red Sandstone' once extended widely beyond its main outcrop over the worn-down surface of the Hercynian mountains. Of such former extension the long Crediton-Hatherleigh tongue of mid-Devon is a clear indication. Further west, a Permian outlier occurs at Babbacombe on the shores of Bideford Bay ; on the south coast such outliers extend to the Plymouth area, and off-shore exploration by direct sounding and seismic prospecting has recently proved [1] a large area of 'New Red' rocks on the Channel floor north of the Eddystone reef. It is equally certain that the Cretaceous cover must have exceeded its present limits ; if we replace the Chalk in imagination above the summit of the Haldon Hills it is clear that it must have passed on to or above the present higher parts of Dartmoor. Linton [2] has recently found reason to suppose that the culminating point, High Willhays (2,046 ft.) may mark approximately the level of the basal plain of the Upper Cretaceous rocks thus projected, and a similar relation may be tentatively inferred in the case of the high Exmoor summits. Between the two upland areas the Cretaceous cover probably rested on Triassic rocks. These considerations indicate that there may well be elements in the

[1] M. N. Hill and W. B. R. King, 'Seismic Prospecting in the English Channel and its geological interpretation' *Quart. Journ. Geol. Soc.* **109** (1953), 1–19

[2] D. L. Linton, 'Problems of Scottish Scenery' *Scot. Geog. Mag.* **67** (1951), 65–85

present landscape which represent the respective 'floors' of the former cover rocks. We should not expect, however, to find such old surfaces in a nearly horizontal attitude, save quite locally. The original base of the Permo-Triassic desert sediments may well have been irregular, but, further, the cover rocks have necessarily suffered later folding and faulting. The Permo-Trias cover is strongly tilted and often faulted at its margins. It survives, for instance, as outliers at 900–950 feet on both sides of the Exe valley north of Tiverton, but beneath the main outcrop barely two miles away, the Permian base is probably below sea-level. We should at least expect both the New Red and Cretaceous covers to record the effects of the early Tertiary movements which raised the Chalk sea floor into land, and the mid-Tertiary or Alpine movements accountable for the strong folding of southeast England. Folding of this date is believed to extend into eastern Somerset and it is highly probable that contemporary faulting occurred farther west. The lacustrine deposits of Bovey Tracey (Upper Oligocene or early Miocene in age) rest in a depression interpreted by Clement Reid [1] as a rift-valley, bounded by faults trending NW–SE. The similar Petrockstow deposits of northwest Devon also rest in a depression which may well mark the continuation of the same zone of faulting. Faulting of similar trend and probably of the same date is conspicuous throughout the peninsula, notably in eastern Cornwall.

It appears, then, that the level 'upland-plains' of Devon and Cornwall cannot be interpreted, as a whole, simply as 'fossil-surfaces' stripped of the later rock-cover. The alternative view relegates these surfaces, as it were, to 'mid-air', regarding them as largely consumed by river erosion in Tertiary times. This was essentially the standpoint adopted by W. M. Davis,[2] who represented the main upland surface of the peninsula as a single peneplain, with certain elements, such as the Dartmoor granite and the more resistant members of the Exmoor sequence, rising above it as monadnocks. This view is readily consistent with the widely accepted conclusion that the higher summits of south-

[1] W. A. E. Ussher and others, 'The Geology of the country around Newton Abbot' (Sheet 339) *Mem. Geol. Surv.* (1913)

[2] W. M. Davis, 'Glacial Erosion in North Wales' *Quart. Journ. Geol. Soc.* 65 (1909), 291

east England between 700–1,000 feet represent a peneplain of Miocene date. It is reasonable to infer that the long period of erosion there represented must have run its course also in South-west England and left a legible mark upon the landscape, such as we might seek to identify with one or more of the high plateau surfaces.

The general course of later investigation proves, however, unfavourable to this interpretation, finding evidence which is held to indicate successive episodes of marine planation. The coastal '400-foot platform' in Cornwall, with a clearly marked but degraded cliff-feature behind it, has long been known and was early regarded as of Pliocene date by Clement Reid by reason of its association with the fossiliferous St Erth beds. Later Barrow [1] recognised the remarkable high-level platforms of the Bodmin area at approximately 750 and 1,000 feet, and by analogy with the 400-foot feature he concluded that these, too, were old 'strand-flats'. Whatever their origin such platforms have necessarily been modified, and locally completely destroyed, by later river dissection. This has obviously been at a minimum along the main peninsular water-parting, and our brief description above suffices to show that the surviving summits fall into distinct groups suggestive of a successsion of platforms. A statistical study by Hollingworth [2] of the frequency of occurrence of summit heights throughout the peninsula fully establishes and extends this conclusion. He confirms the presence of the surfaces earlier recognised at about 400 and 1,000 feet and finds strong indications of others at 520–540 feet and 620–640 feet. Between 740–940 feet two distinct platforms are in evidence, and there are indications of a surface at about 1,150 feet, above the 1,000-foot platform.

Further confirmation comes from two field studies by J. F. N. Green in east Devon [3] and W. G. V. Balchin [4] on Exmoor. Both these workers have mapped the surviving portions of a series of

[1] G. Barrow, 'The High Level Platforms of Bodmin Moor' *Quart. Journ. Geol. Soc.* **64** (1908), 384–400

[2] S. E. Hollingworth, 'The Recognition and Correlation of High Level Erosion Surfaces in Britain' *Quart. Journ. Geol. Soc.* **94** (1938), 55–84

[3] J. F. N. Green, 'The High Platforms of East Devon' *Proc. Geol. Assoc.* **52** (1941), 36–52

[4] W. G. V. Balchin, 'The Erosion Surfaces of Exmoor and Adjacent Areas' *Geog. Journ.* **118** (1952), 453–476

117

platforms and the bluffs which mark their upper limits, regarded as degraded coastline features. Their results are significantly accordant though they differ in the number of platforms distinguished. In east Devon Green recognises a limited area of a 1,000-foot platform on the summit of the Blackdown Hills, succeeded in order southwards by a series of platforms of which the ranges of height, from the base of the upper boundary bluff to the forward edge, are as follows : 920–750 ft., 690–620 ft., 595–545 ft., 530–520 ft., 505–460 ft. In similar terms, Balchin's sequence of platforms is as follows : 1,225–1,000 ft., 925–800 ft., 825–750 ft., 675–500 ft. Both authors find local evidence of the 400-foot platform, and Balchin recognises a lower surface at about 200 feet.

The problems of the origin and correlation of such platforms have occasioned much discussion [1] which we cannot here pursue. The conclusion most naturally emerging from the evidence is evident enough. Since the elevation and spacing of the platforms is significantly accordant over long distances, the movements indicated are uniform or 'eustatic' downward movements of sea-level, rather than successive uplifts of the land which could hardly have been unaccompanied by some element of warping. With the wider implications of the theory we are not here concerned. It is certainly entitled to rank as a working hypothesis. The southwestern peninsula is one of the regions which afford cogent evidence for the hypothesis and in which its acceptance or rejection greatly affects our view of general physiographic evolution, and especially of drainage development.

The Relation of the Drainage to the Erosion Surfaces

Let us therefore summarise, so far as present knowledge permits, the light thrown by the drainage pattern on the age and origin of the erosion surfaces (Fig. 2). The problem involved may best be seen if we consider, in turn, the western, central, and eastern sections of the peninsula. The first includes Cornwall and those parts of Devonshire falling within the Tamar and

---

[1] S. W. Wooldridge, 'The Upland Plains of Britain : their origin and geographical significance' *Advancement of Science* 7 (1950), 162–75. See page 124 of the present volume.

Tavy basins. Here, if we discount the later northward diversion of parts of the drainage, we have a simple pattern implying an initial tilt towards the south or southeast. The most probable date for this movement is mid-Tertiary (Alpine) and it is likely to have been accompanied by faulting. The form of the Tamar basin suggests that it may well mark a former tectonic depression extending across the peninsula, its original margins defined by warping, faulting, or both. If we assume that the course of sub-aerial development, following this movement, was unbroken by any major marine incursion, we can identify the 1,000-foot platform with a peneplain marking the conclusion of a first cycle of erosion. If, on the other hand, this platform and its successors between 920–750 feet are regarded as marine, we should be led to infer that the drainage resumed or at least imitated its former lines as the sea retreated. The relics of these higher platforms within and west of the Tamar basin are limited in extent ; the 920-foot bluff is known in the Bodmin Moor area but its course as a whole is not yet clear. It is impossible, indeed, on present evidence to reconstruct the position of the supposed marine platforms which may well have been cut partly in Trias and since completely destroyed. Further detailed study of the area is in progress. At first sight at least the pattern of the drainage on both flanks of the Tamar basin hardly suggests that it has been extended, stage by stage, across a series of emergent marine platforms with the streams 'normal', as they would be expected to be, to the line of the supposed coastlines.

No such difficulty is encountered in the case of the lower platforms. Throughout Cornwall the drainage pattern is closely consistent with 'normal' extension across the 400-foot platform. Before the emergence the Land's End and Carnmenellis granite masses must have been islands, and the western end of the St Austell mass was the 'Land's End' of the period. Although it has not yet been traced in detail there is also good evidence of a platform between 700 and 550 feet throughout southern Cornwall and the drainage pattern is normally adjusted to it.

Coming now to the central section of the peninsula between the eastern limit of the Tamar basin and the line of the Exe, we are confronted with very different conditions. The drainage pattern shows a relatively strong development of east-west lines, and though some of these mark the picking out of the structural grain of the country by subsequent streams, there are significant evidences of the survival of an older drainage system flowing eastwards. This was early perceived by A. W. Clayden and A. J. Jukes-Browne, while in his last paper J. F. N. Green [1] extended and modified their hypothesis. It is possible to regard the eastward-flowing streams of Exmoor, central Devon, and Dartmoor as relics of an early Tertiary drainage system flowing originally on the Chalk, tilted eastwards towards the Eocene sea in Dorset. A later southerly tilt, no doubt contemporary with that of Cornwall, is also in evidence, but this would not be expected to displace the eastward-flowing rivers, by then entrenched in the Palaeozoic undermass. It is clear, in fact, that it did not do so in the Dartmoor and Exmoor regions, though Green has ingeniously argued that it was responsible for the suppression of the northward-flowing tributaries of the Dart on Dartmoor. The drainage pattern of central Devon is more complicated and difficult to interpret, but following Green's line of thought we might sustain the hypothesis that the three main groups of southward-flowing streams, on the flanks of Exmoor, parts of Dartmoor, and in the southern coastal zone were strengthened by the southerly tilt, but initiated as left-bank tributaries of eastward-flowing trunks. This might be taken to imply that there was no major marine incursion in the area, except that of the 400-foot sea in the south.

At the line of the Exe, however, we enter country where the dominant drainage direction is again southerly, and eastward-flowing elements, apart from the subsequent Tone, are almost completely absent. This gives strong presumptive evidence of a marine submergence which obliterated the older eastward-flowing drainage. The high surface or surfaces across which the Exe must once have flowed south of Exmoor were obviously cut largely in Permian rocks ; superimposition from the Permo-Trias cover sufficiently explains the transection by the river of high ground both north and south of Tiverton and again north of Exeter. Farther north, the Quarme-Exe headstream together with those of the Haddon, Batherm, and Tone form a sub-parallel family of streams, of youthful aspect, entrenched below

[1] J. F. N. Green, 'The History of the River Dart' *Proc. Geol. Assoc.* **60** (1949), 105–124

the surface of Balchin's 1,225-foot surface. Professor O. T. Jones [1] has objected that, farther west, the drainage is not 'normal' to Balchin's suggested coastlines. The objection, however, does not survive closer scrutiny, and south of the Brendon Hills the normal relationship is undoubted and gives positive support to the marine origin of the summit-plain. Without a marine incursion at this level it is difficult, if not impossible, to explain the suppression of the older eastward-flowing drainage. The ascription of so radical a change to 'cross-tilting' would be highly implausible and flagrantly counter to established principles of drainage development.

In the country west of the Exe farther south the evidence of marine interruption of drainage development is less decisive, or at least incomplete. The upper sections of the Teign and the Dart, regarded as the headstreams of former right-bank tributaries of the Exe, must have maintained their eastward flow well beyond the high ground of the Dartmoor granite and retained it until a relatively late date, as is manifest from the abandoned valley of the Dart east of Ashburton. We are left to conclude either that the higher level marine platforms were not developed in this part of the region or, alternatively, if they were in fact developed as Green contended, they were so disposed as to permit the eastward extension of the Teign and the Dart without sharp change of direction. Such a reconstruction may yet prove possible, but our present evidence does not suffice to justify it.

East of the Exe no such difficulties exist, for the dominant southward-flowing drainage is consistent in its lines with the presence of the series of marine platforms claimed by Green. [2] Even if the marine origin of these surfaces were denied, it is important to note that they are, beyond question, Tertiary surfaces, cut in Cretaceous rocks and the absence of corresponding features in the more westerly Palaeozoic terrain would be quite unaccountable.

The conclusion therefore to which our brief survey points is generally favourable to the marine hypothesis, though local discrepancies remain to be explained, and there is some evidence

[1] In discussion with W. G. V. Balchin, *Geog. Journ.* **118** (1952), 473–4
[2] J. F. N. Green, *op. cit.* (1949)

that parts of the central section of the peninsula may have escaped submergence at the higher levels. The evidence certainly does not support the supposition that the land-forms of the area have been exhumed largely intact from beneath a Permo-Triassic blanket. Nevertheless the former wider extension of the cover is a factor to be reckoned with. The high scarp-like feature which bounds Exmoor on the west and south is strongly suggestive of the front of a desert upland rising above a 'pediment' plain, and the same may possibly be true of parts of the steep margins of Dartmoor. This would imply that parts of the plateau surface at 800–1,000 feet mark essentially surviving portions of old desert pediments modified in varying degree by sub-aerial or marine action. However, this may be, it seems highly probable that the coasts of the peninsula, particularly on the north, retain in their main features legacies of sub-Triassic geography. It is exceedingly difficult to explain the consistent northward diversion of formerly southward-flowing drainage without assuming that the Bristol Channel depression dates from early Mesozoic times and was thickly filled with Triassic and later deposits. The proximate cause of the various episodes of river capture has been the rejuvenation following the withdrawal of the sea from the 400-foot coast, but this leaves unanswered the prior question of how the sea gained access to the area. A likely hypothesis seems to be the earlier excavation of the filling of the depression by a major river flowing west, of which the left-hand tributaries, excavating readily in the Trias, quickly reached the vicinity of the present north coast, marking roughly the margins of the old depression. Here they would have begun to compete with the southward-flowing drainage and prepared the way for the ultimate reversal of part of the latter, consummated only by the 400-foot submergence and the ensuing uplift.

It may well seem that the physiographic story we have sought to outline is at once too complex and too incomplete readily to serve the needs of elementary teaching. Let us beware, however, of the charge so often brought against us that we are interested only in the easy and the obvious. It is neither possible nor desirable to present, even at sixth-form or first-year University level, a simple, neat, and final solution of the physiography of the Southwest, pre-digested and suitable for regurgitation. But this need not impede our own recognition that the physical no less than the cultural landscape has a past which informs its present. The area abounds in beautiful examples of physical phenomena of world-wide significance, but the examples are drawn from and contribute to a record of continuous evolution in which the hills are veritably shadows and 'flow from form to form'. Only in the light of the tale as a whole, as we have tried here briefly and incompletely to tell it, can the significance of such examples be fitly seen, and only so can we isolate the gaps in our knowledge which are a challenge to future research.

9

The Upland Plains of Britain

Their Origin and Geographical Significance

I

MY AIM in this address is to survey some of the problems of the later stages of landscape evolution in Britain and more particularly the evidence afforded by the various well-marked erosion surfaces developed on rocks of widely varying ages. My subject, therefore, falls within the strict and proper purview of geomorphology, and during the last thirty years work upon it by both geographers and geologists has been very actively pursued in Britain. The time seems ripe for a review of such work ; if it cannot provide answers to all the problems arising, it may at least succeed in asking some of the right questions and thus guide the direction of further research.

If in pursuing this subject we find ourselves cultivating the common borderland of geology and geography I find no need to apologise for that here. It is clearly part of the tradition of the British Association that inter-sectional interests and issues should be discussed. For myself, I do not doubt the close relevance of the physiographic findings to the general tasks of geographical interpretation. That relevance has been evident to a large number of geographers and to it we certainly owe the initiation of the 'Terrace-Commission' of the International Geographical Union which has been at work since 1926. The instinct which prompted and still maintains the investigation is sound enough. One can draw no tidy division, armed with 'keep out' notices, between the fields and interests of geologists and physical geographers. I have no desire to enter upon one of those methodological disputations of which we, as geographers, are perhaps inordinately fond, and I will content myself with affirming that, if the study of the shape and shaping of the physical landscape is ever expelled, through prejudice or ignorance, from the ambit of geography, our subject will have lost much of its historical birthright, including even the right to its time-honoured name. Welcoming cordially as I do the marginal growth of our subject into the fields of study of 'social man', I flatly rebut the suggestion that this alone is 'real geography'. Like geomorphology itself, it is part and only part of a whole. As I see it, one of the emphases at present needed in British geography is a reminder of the virtue of keeping both feet firmly on the ground—'the solid ground of Nature', if not quite in the sense in which Wordsworth first used this famous phrase. For 'ground' you may read 'land' or, if you insist, 'landscape'. But let us at least be clear that landscape involves more than the so-called cultural landscape. However important a rôle we ascribe to the latter, we shall make but little of its study without much closer scrutiny of its physical foundations than is now fashionable in some quarters.

2

By a combination of physical chance and human perversity the Tertiary era has almost become a neglected 'Dark Age' in the geological history of Britain. Despite, or perhaps because of, the general wealth of our geological record in these Islands there is as yet no adequate interpretation or even description of their geomorphology. The geologist deals convincingly and at length with what has been called palaeogeography, but the geographer turning to these findings for the antecedents of the features and phenomena of his study is too often handed a stone in place of the relevant bread. For palaeogeography is necessarily and inevitably an affair, primarily, of long-past sea-bottoms and of coastlines tentatively located. The evolution of the Tertiary land surfaces, not less a proper inquiry in palaeogeography, presents quite different problems, amenable to attack only by the methods of geomorphology.

These problems are rendered difficult of solution by the limited and fragmentary character of our Tertiary sedimentary record. This default of evidence, taken with the fact that many of the British uplands show manifest structural lineaments of ancient date, and present seemingly simple relations to the lowlands, puts a premium on over-simple explanations of the physical features of Britain. They are coloured, too, by the widely received conclusion that the Upper Cretaceous transgression left little of the British area unsubmerged. The widely spreading sheet of Chalk, or equivalent deposits, uplifted in early Tertiary times, becomes as it were the *terminus a quo* of British landscape evolution, and, to a first approximation, all that seems necessary is to conceive of the stripping of this Cretaceous mantle from upland and lowland alike, thus allowing the older structural patterns to show through. More than a century of detailed work on the succession and structure of Britain enables us to give graphic and circumstantial accounts of the antecedents, climax, and sequel of the great episodes of Caledonian and Hercynian mountain-making. The faulting which accompanied or succeeded the latter is still plainly legible in the marquetry of the British landscape, and we can trace, in gratifying detail, the later story of the Triassic deserts and the Jurassic and Lower Cretaceous seas, until the great Upper Cretaceous transgression closes and, as it were, seals off the account. It is, naturally enough, such episodes as these which have always taken pride of place in the tale of British palaeogeography as written by pioneers like Edward Hull [1] and A. J. Jukes-Browne [2] or in broader outline by geographers seeking an evolutionary background to their canvas. The later parts of the record, excepting only the Alpine folding in the South Country, receive much scantier treatment in the standard accounts, though they are obviously much more important for the student of present surface forms. Jukes-Browne, indeed, made a brave and ingenious attempt to trace the progress of Tertiary denudation ; but he was necessarily denied much of the evidence now available and many of his conclusions are no longer tenable. Later, Professor L. J. Wills [3] very clearly and fairly summarised the more recent work done in southeast England, but he did not find occasion to deal with the landscapes of Devon and Cornwall which probably hold not a few of the keys to the interpretation of our Tertiary land surfaces.

In the light of the evidence gained during the last thirty years it is now time that geologists and physical geographers faced

[1] E. Hull, *Contributions to the Physical History of the British Isles* (Stanford, 1882)

[2] A. J. Jukes-Browne. *The Building of the British Isles* (Stanford, 4th ed., 1922)

[3] L. J. Wills, *The Physiographical Evolution of Britain* (E. Arnold, 1929)

the fact that one cannot, in any real sense, explain, say, the Welsh mountain mass in terms of Caledonian orogeny, which was a phase not so much in its history but its pre-history. Still less are the land forms of Devon and Cornwall at all closely related to the Hercynian folding ; they bear the plainest impress of Tertiary faulting and late Tertiary planation. To elide, as we constantly tend to do, some sixty-five million years of geological process, accountable to Tertiary time, is to import a sort of 'antiquarian' squint into our physiographic interpretations. Since Tertiary deposits obtrude so little, and over much of Britain even Cretaceous rocks are missing between the solid basement and the Pleistocene drifts, physiographic accounts perform a sort of leap-frog antic and pass in one breath from tracing the topographic expression of ancient structures to the story of glaciation. Since no deposits intervene, it has been widely assumed that nothing can safely be said of the vast 'lost interval', or that it is a field, at best, for aimless and profitless speculation. It is this great lost interval in the geological story, and a corresponding area of ignorance in the interpretation of the geographer's landscapes, that the methods of modern geomorphology are fitted to fill. To ignore them and to leave the lost interval intact is as if we sought the roots of some modern politico-social problem in minute and careful studies of the manorial system, ignoring the immediately precedent stages of history. Or it may afford a closer parallel to our case if we imagine an archaeologist reconstructing the original form and stages of decay of some major building of the past. Let it be assumed that it was fashioned of local building stone. Even so the grain or other internal structures brought out by weathering in the building blocks would not form valid parts of the archaeologist's reconstruction. The plan and fashion of the original building and the successive phases of its demise would reflect quite other processes than those which formed or weathered the stone. We must not, of course, press this analogy too far, for, in our case, the great principle of geological continuity is at stake. The later are of one piece with the earlier episodes as parts of a continuous physical evolution. But if, as Professor O. T. Jones recently noted,[1]

[1] O. T. Jones, 'The Structural History of England and Wales' *Rep. Int. Geol. Cong.* Part I (1950), p. 216

the greater part of Britain has been structurally dead since the close of the Hercynian movements, it has none the less been morphologically very much alive. If the structural geologist has little or nothing to say of the vast periods which appear to him quiescent, the geographer must needs turn to geomorphology for guidance in his own analyses.

3

In broad terms there are two methods of tackling the problem of the lost interval. The first consists in the analysis of the drainage system and its evolution. It was first applied in this country by W. M. Davis, followed by Cowper Reed, Buckman, Strahan, and Lake, and, in more recent years, notably by Linton and Bremner. I shall not attempt to deal here with the method and its findings. Arguments of considerable elegance can be based upon it, but they tend, of themselves, to be indecisive unless definite stages in chronological sequence can be established. This the second method supplies, for it aims at the recognition of past base-levels of erosion whether represented by submarine or sub-aerial surfaces. Associated with it is the analysis of composite or cyclic river profiles first applied in this country by O. T. Jones in 1924 and since developed by the late J. F. N. Green. I shall not discuss it at length since it is definitely ancillary to the study of erosion surfaces. Miller has shown the wide margins of error which characterise the method of base-levels determined by simple extrapolation of river curves. Without the check imposed by erosion surfaces and river terraces, reconstruction of past river profiles could hardly yield dependable or definitive results. Far greater importance must be assigned to plateau-like erosion surfaces, for these are actual geographical features which can be mapped with a reasonable approach to accuracy and traced for considerable distances. Only the preliminary question of nomenclature need detain us before we review on a regional basis the considerable body of evidence now to hand.

4

It is usual to refer to the features in question as 'erosion surfaces' or 'erosion platforms', and in ordinary working parlance these terms are unobjectionable and too well entrenched to be easily replaced. Yet neither term is free from ambiguity, and since geographers are showing timely signs of a certain tenderness of conscience in the matter of terminology it will not be amiss to employ a little circumspection. It is clear that there are many erosion surfaces, e.g. valley sides and cliff faces, which are in no sense platforms. The term 'platform' is graphic and self-explanatory, and in some cases apt enough, yet the purist may insist that a platform should be flat or nearly so. In fact, most of the surfaces in question are appreciably inclined, whether by virtue of original slope, subsequent dissection, or tectonic warping. Moreover some authors use the term 'structural platform' where the top of a resistant master-stratum is locally stripped of a conformable overburden. Such a platform is evidently, in some sense, an erosion surface, though not generally in the sense in which 'the platform school' use the term. It is, of course, obvious enough that the significant features for our purpose are surfaces of low relief inferred from accordant summit levels or preserved as flats truncating or bevelling the structure. It is to be noted that we cannot use a genetic nomenclature without begging the very question we are required to solve. Some of the surfaces are no doubt peneplains in the Davisian sense, but unless that term is given a grossly widened and weakened connotation it cannot be applied to them all. I do not favour the use of the term marine peneplain. Nor is this all. Extensive surfaces of low relief bevelling structure may be neither peneplains nor strand-flats, but the produce of lateral river corrosion, arid planation, or advanced corrie sculpture. There is the further difficulty that a surface may be 'fossil'—i.e. an exhumed plane of unconformity, which commonly leaves us in a further doubt whether its original fashioning was submarine or sub-aerial, or in part both. To press these difficulties to the limit would no doubt be pedantic, yet they are real enough if we seek a definitive and unambiguous glossary definition of erosion surfaces and platforms. I shall not forgo the use of these simple and essentially commonsense terms in reviewing the British evidence, but my title revives a practice of the Geological Survey in Devon and Cornwall and notably of Mr Henry Dewey. Generically, as geographical features, the chief members of the series of surfaces are 'upland-plains'. In using this term one begs no question at all and merely states an observed fact. Even those who are sceptical of the reality of erosion surfaces must surely grant that upland plains are a conspicuous feature of the physiography of Britain. There is the further advantage that the term is applicable to composite surfaces, peneplains retaining an appreciable relief, or a series of closely spaced platforms difficult or impossible to separate. I will not pursue further what might easily become an irritating discussion. I submit, merely, that to speak of the 'upland plains' of Britain conveys at once more and less than the 'platforms' of Britain. If anyone thinks that I might as well have said the plateaux of Britain, I will ask leave to express my personal disagreement without further argument.

5

I must attempt now the far from easy task of summarising the evidence at present available under regional headings.

If I begin in southeast England it is not because this is the region I personally know best, but because here Tertiary deposits, and folding of known age, put our problem in its correct time setting. The cardinal feature is the Pliocene platform developed on the Chalk outcrops flanking the London Basin and either bevelling the escarpment or forming a distinct bench on the dip-slopes.[1] The deposits which rest on it, originally ascribed to the Older Pliocene have since been shown to yield a Newer Pliocene (Red Crag) fauna.[2] The typical elevation of the platform and one faithfully maintained is about 600 feet, but it ranges downwards to 550 feet or perhaps slightly lower, while what has been taken as a degraded cliff-line feature at its back comes on somewhat below 700 feet. Where the platform adjoins higher ground, as in the North Downs of Surrey or Kent, or in the Chiltern Hills, this ground shows distinctive features both in its drainage patterns and soil cover which are consistent with the conclusion that it was not submerged at this time. On the platform the drainage lines could not arise till the Pliocene sea floor was uplifted, and they must have been ultimately superimposed

[1] S. W. Wooldridge, *Proc. Geol. Assoc.* **38** (1927), 49
[2] C. P. Chatwin, *Summ. Prog.* (*Mem. Geol. Surv.*, 1927), 154. Also *Summ. Prog.* (1930), 2

from the cover of sand and shingle. With the help of this clue the coastline was traced by Professor Linton and myself southwestwards from near Hitchin to the neighbourhood of Dorchester.[1] It occurs also in a dissected condition on the South Downs, but we concluded that much of the central Wealden area remained unsubmerged.

Within the Chalk areas the recognition of the Pliocene platform defines also not only the higher tracts but the exhumed sub-Eocene surface. The former areas have been regarded as surviving relics of the Miocene or Mid-Tertiary peneplain. The latter forms a distinctive facet on the lower part of the Chalk dip-slope. Only rarely as in the western Chilterns does it form a true upland plain. Its existence reminds us that a lengthy period of erosion ran its course before the earliest Eocene deposits were laid down : there is evidence that the Chalk cover was locally breached by early Eocene times. Though the sub-Eocene surface was locally trimmed by the waves of the Eocene sea it was essentially a peneplain, and the possibility that it may prove recognisable in northern or western Britain must be borne in mind.

Between the Pliocene platform and the 100 feet or Boyn Hill terrace of the Thames, a number of gravel-covered valley-floor plains can be distinguished in the area north and west of London, but at two levels only is there evidence of widespread planation. What has sometimes been called the 400-foot platform dominates the form of the country north of London.[2] It forms the summit plain of Middlesex, south Hertfordshire, and much of Essex, though in the latter area it nowhere attains 400 feet, dropping eastwards to 350 feet or below. I formerly regarded this feature as a river plain and there can be no doubt that the ancestors of the Thames and its southern tributaries flowed across it and left deposits upon it ; but I am now inclined to revert to the view of George Barrow that the greater part of the deposits (pebble gravel) are marine, though later in age than those of the main Pliocene platform. Evidence of a still-stand

[1] S. W. Wooldridge and D. L. Linton, *Journ. Geomorph.* **1** (1938), 41, and in 'Structure, Surface and Drainage in Southeast England' *Inst. Brit. Geog.*, Publication No. 10 (1939)

[2] See S. W. Wooldridge, *Proc. Geol. Assoc.* **38** (1927), 49, and G. Barrow, *Proc. Geol. Assoc.* **30** (1919), 36

at about the same level is found in the northern and central Weald though there is here no suggestion of marine planation.

At a lower level what I originally termed the 200-foot platform is more widely represented.[1] It shows every sign of being a surface of sub-aerial planation and was formed before the major glaciation of the area. As we shall see, a comparable feature has since been widely recognised in Britain : in part at least it corresponds to the Ambersham stage of J. F. N. Green. I must plead guilty to having tentatively correlated this feature with the Milazzian marine stage of the Mediterranean Pleistocene. I have often since been rather aghast at my carefree rashness in so doing, but the suggestion still agrees in broad terms with the local evidence, and I note that Professor Zeuner accepts the correlation.

A recent contribution to the denudation chronology of southeast England is the work of Mr B. W. Sparks on the seaward slope of the South Downs.[2] Here he recognises no fewer than eight dissected shelves between the Pliocene platform and the well-known Pleistocene raised beach of Goodwood. They range in height from 475 feet to 180 feet with an average interval of about 42 feet. The best developed of the features are those at 475 feet and 430 feet respectively, while those at 345 feet and 180 feet are relatively feeble. None of the features attains the magnitude of an upland plain : they are narrow facets on the dip-slope. To those who find the number of stages embarrassingly large I will say only that the record deduced by Sparks is wholly consonant with that of the area as a whole ; and if we accept the fact of marine planation at the 600 feet (Pliocene) and 100 feet (Goodwood) levels, it indicates a consistence and expectable sequence of coastal evolution between those dates.

6

Passing now westwards we find that southern Hampshire, despite its impressive terrace sequence, has little to offer us at the higher levels, though the New Forest summits east of Fordingbridge may retain isolated fragments of a 400-foot plat-

---

[1] S. W. Wooldridge, *Proc. Geol. Assoc.* **39** (1928), 1
[2] B. W. Sparks, *Proc. Geol. Assoc.* **60** (1949), 163

form. In tracing what we regarded as the main Pliocene shore-line across Wessex to the Dorset coast, Professor Linton and I fixed upon the break of slope above the Hardy monument near Dorchester as the back of the 600-foot platform, and this still appears to be a reasonable inference. Though the intervening ground awaits detailed scrutiny, less than twenty miles westwards, across the Devon boundary, the sequence of high level platforms is strongly resumed. Here in the familiar hinterland of Sidmouth, Seaton, and Lyme Regis, where the flat-topped mesa-like masses of Upper Greensand rise above the Triassic basement, one is readily led into the facile assumption that the summit plain is a structural platform—i.e. the stripped surface of the Upper Greensand. This is not so. The surfaces transgress the structure locally, quite apart from the fact that they are developed to the east on the Chalk and the west on Palaeozoic rocks. J. F. N. Green[1] distinguishes a remnant of a 1,000-foot platform on the crest of the Cretaceous escarpment south of Taunton. Below it the broad plateau between 920 feet and 750 feet is identified with Barrow's Bodmin Moor platform in Cornwall, while nearer the coast, platforms appear with backs, in descending order at 690 feet, 595 feet, 530 feet, and 505 feet. High valley-side flats in the Axe valley lie at about 440 feet and may be regarded as grading to a shoreline at a slightly lower level not now pre-served in the area. Green regarded all the platforms from 505 feet upwards as marine. They present indeed the unmistakable aspect of strand-flats, and the record of shingle on the Bodmin Moor surface near Otterford affords further support in this case. Though I am sure the last word has not been said on this ground, and it remains for us to trace Green's features eastwards through Dorset I entirely accept his broad conclusions and regard them as of the highest importance.

Let us then turn northwards for a moment to the Bristol district where upland plains cut in Carboniferous and younger rocks are plainly in evidence. As regards the former as seen, for example, at Clifton and Durdham Downs it was for long cus-tomary to assign the planation to the work of the Rhaetic or Liassic seas. This is a good example of the recurrent tendency to interpret our British platforms as fossil surfaces, thus gravely

---

[1] J. F. N. Green, *Proc. Geol. Assoc.* **52** (1941), 36

underestimating the duration, and what may be termed the planation-potential, of Tertiary times. The theory is highly implausible here since the Lias has itself been subject to well-marked folding on east-west axes, probably of Mid-Tertiary date. A. E. Trueman stated the case for late Tertiary planation [1] recognising erosion platforms at 200–300 feet, 400–450 feet, and 550–600 feet. while a higher surface at 750–800 feet is prominent in the Mendips, on Dundry Hill, and on the high summits of the Cotswolds north of Bath. Four years war-time evacuation at the University of Bristol offered me some limited opportunities of following Trueman's lead in this area, but I will confine myself here to two observations. The Durdham Downs platform, so prominent a feature of the Bristol scene, must in my view be regarded as including a representative or continuation of the 400-foot surface, though no clear-cut upper limit against higher ground can be fixed in the immediate environs of Bristol. The general level of the higher parts of the platform is 350 feet, and though it ranges down to the lower levels the same is true, as we shall see, in many parts of both Cornwall and South Wales. That the surface is at least post-Liassic was clearly proved by a large bomb crater in Westbury Park on the eastern side of the Downs : this revealed Lias levelled flush with the Downs surface cut in Carboniferous Limestone.

The other feature which greatly impressed me was the widespread planation at between 550 and 600 feet in the Oolitic country south of Bath. North of the Avon valley the great bastion of Lansdown rises to 780 feet, and viewing it northwards across the lower surface, I was inevitably reminded of the feature I had mapped in the London area behind the Pliocene shelf. The nearest area of Pliocene planation surveyed by Linton and myself lies some twelve to fifteen miles to the southeast on Salisbury Plain ; if the same sea covered both areas the only direct continuity between them seems likely to have been *via* the breach in the high Chalk country near Warminster. Flint is widely spread over the high ground south and east of Bath, and its distribution merits most careful mapping.

[1] A. E. Trueman, *Proc. Bristol Nat. Soc.* **8** (1938), 402

I have considered the East Devon and Bristol areas at this point of the argument since they form a link between southeast England and the southwest peninsula, where striking evidence of planation has long been known. As early as 1890 Clement Reid associated the famous coastal platform at 430 feet with the St Erth beds, assigned by him to the Older Pliocene.[1] Not until the Geological re-survey of the area reached the Bodmin mass were considerable relics of higher platforms encountered ; but in 1908 [2] George Barrow described the 1,000-foot and 800-foot (Bodmin Moor) platforms here, and both were later identified on the flanks of Dartmoor. Barrow regarded both as marine and they certainly bear all the aspect of strand-flats, but Clement Reid, speaking later of the Dartmoor area, was inclined to invoke a sub-aerial origin.

There has been no general review of the Cornwall and Devon evidence since the completion of the re-survey nearly forty years ago, though some local studies have been made. What is evidently required is a wide extension of the careful and detailed type of work which my colleague, Mr Balchin, is communicating at this meeting. Meanwhile we may note that the whole of the tentative chronology which has been generally accepted, ascribing the 400-foot platform to the Older Pliocene and the higher surfaces to the Miocene, rests on Clement Reid's verdict as to the age and depth of accumulation of the St Erth beds. I suspect that this chronology is badly in error. The St Erth beds lie at about 100 feet O.D. and were accumulated, according to Reid, in some forty fathoms of water. This would give a sea level at about 340 feet, and the link of the low-lying fossiliferous sediments with the much higher strand-flat was thus accomplished. In the matter of age no great difficulty existed while the 600-foot platform of southeast England was regarded as of Older Pliocene age. But since we now know that the deposits near London yield a Newer Pliocene fauna, the discrepancy of level is serious and would imply warping, of which neither region nor the intervening areas shows any clear sign. Morphological evidence

[1] C. Reid, 'The Pliocene Rocks of Britain' *Mem. Geol. Surv.* (1896)
[2] G. Barrow, *Quart. Journ. Geol. Soc.* **64** (1908), 384

would in fact suggest a much later age for the 400-foot platform. I believe it may well prove to be not older than latest Pliocene in age. This would bring it into line with the corresponding feature in the London district, but would imply either the divorce of the platform from the St Erth beds or a revised view of the age of their fauna. It would also imply that a 600-foot surface should be recognisable in the southwest. In fact both Balchin [1] and Gullick [2] have found local indications of such a surface in Cornwall, and Balchin has also located it on Exmoor with its back at the usual level of between 650 and 700 feet. It is imperative that this surface should be carefully traced. A glance at the map reveals many areas which might include it, but there is also a strong suggestion of general sub-aerial grading to or towards the 400-feet coastline, in the course of which initial surfaces between 500 and 1,000 feet have been modified or destroyed. Such a grading process, of which Miller has found evidence in South Wales and elsewhere, was of course inevitable. It is clearly shown over many areas of the 400-foot platform itself, notably on the south of the peninsula between St Austell and Dartmouth. The sub-aerial imprint on older platforms is necessarily even more strongly marked, but I believe that it will be possible to distinguish residual areas of marine planation.

The existence of a lower platform at about 200 feet in the area has also been claimed by several workers and there is, indeed, clear evidence of a still-stand at this level, though the surface often merges imperceptibly upwards into the 400-foot strand-flat.

8

It thus seems possible to establish a base or reference line more than 300 miles long from Kent to Cornwall along which the platform sequence is everywhere broadly consistent, at least at the lower levels. It is a highly remarkable and significant fact that we can now run another line from Cornwall to Cumberland along which entirely similar evidence is found.

I can deal more briefly with this area since it is well covered by Miller's review of 1938 [3] and the important statistical analysis,

[1] W. G. V. Balchin, *Geog. Journ.* 90 (1937), 52
[2] C. F. Gullick, *Trans. Roy. Geol. Soc. Cornwall* 16 (1936), 380
[3] A. A. Miller, *Proc. Yorks. Geol. Soc.* 24 (1938), 31

based on altimetric frequency curves, published by Hollingworth in the same year.[1]   In South Wales the work of Miss Goskar, Dr Trueman, and Professor George revealed a succession of platforms, up to at least 600 feet, closely comparable with that of Cornwall.   Professor Miller's work on the area emphasises the probability of a considerable element of later sub-aerial grading or merging of the surfaces.   Miller then pressed the inquiry northwards along the coasts of Ireland and Wales to Cheshire, Lancashire, and the Lake District, where he incorporated the results of the work by Hollingworth.   He found recurrent evidence of marine still-stands at about 800, 450, and 200 feet.   From the 200-foot platform, which is widespread, a sloping surface or a succession of plateaux leads up to a presumed cliff-line at about 450 feet.   Above this a similar surface, locally showing discontinuous cliffing, but often smoothly sloped, ascends to 800 feet, the flatter areas being generally best developed above 600 feet.   Miller has also detected both coastline features in southern Ireland.   Hollingworth's analysis of the seaward slopes of the Lake District revealed six main levels of planation below 1,100 feet (320, 430, 570, 730, 820, and 1,070 feet) and three subsidiary levels at 520, 620, and 920 feet respectively.[2]   A more recent study by E. H. Brown [3] of the hinterland of Aberystwyth claims evidence over the same range of height for major planation stages at 1,000–1,100 feet and 500–675 feet—the main coastal platform.   Four minor intervening stages occur with upper levels at about 1,000, 875, 825, and 750 feet respectively, while extrapolation of river profiles leads him to infer a series of lower sea levels at about 407, 326, and 233 feet.

9

I shall not attempt to pursue this review in detail into Scotland where less evidence is at present to hand, and my ignorance of much of the ground prevents my commenting usefully upon it.   We may note, however, that quartzite-flint gravels at an elevation of 350–400 feet were described as long ago as 1921

[1] S. E. Hollingworth, *Quart. Journ. Geol. Soc.* **114** (1938), 55
[2] S. E. Hollingworth, *Proc. Yorks. Geol. Soc.* **23** (1936), 159
[3] E. H. Brown, *Inst. Brit. Geog.*, Publication No. 16 (1950), 49

in the Buchan district of Aberdeenshire and a Pliocene age has been tentatively assigned to them.[1] In the Central Lowlands Ogilvie has distinguished his Higher and Lower Lowland peneplains.[2] The lower feature is widely drift-covered and is an undulating surface which hardly suggests marine planation ; but the Higher Peneplain, typically developed between 500 and 750 feet might well prove to include one or more strand-flat features.

10

Our review has omitted the eastern coastal areas south of the Border, since the evidence is at present slight and inevitably suspect. The attitude of the Pliocene base in East Anglia and the Low Countries, and the great thickness of Pleistocene deposits in the latter area, show that down-warping of the North Sea depression must have been active during and after Pliocene time. It is clear that the basin is in essence geo-synclinal in its nature, and its growth began in Lower Cretaceous times if not earlier. It is a reasonable presumption that our eastern coastal areas have been affected by this warping. In the south this is beyond question ; and I believe one can locate the effective margin of the North Sea depression in mid-Essex near Chelmsford. For many miles northwards we have no likely source of evidence on this point, though it is obviously desirable that the morphology of the Lincolnshire and Yorkshire Wolds should be closely examined. Versey has examined possible relics of Tertiary deposits on the Yorkshire Chalk.[3] It is significant that his application of Hollingworth's method of altimetric analysis to the Cleveland area fails, on the whole, to show clear evidence of planation at the usual levels. It seems possible that there may be some planation at and below 800 feet in the coastal zone between Saltburn and Scarborough, but in default of a demonstrable sequence of platforms no correlation is at present possible. Hollingworth included the Cheviot area in his statistical survey,

---

[1] H. H. Read, 'The Geology of the Country around Banff, Huntly and Turriff' chap. 11, *Mem. Geol. Surv.* (1923)
[2] A. G. Ogilvie, *Great Britain* (C.U.P., 1937), p. 425
[3] H. C. Versey, *Proc. Yorks. Geol. Soc.* **23** (1937), 302

finding evidence of planation at 560, 760, 920, and 1,070 feet, but not at 400 feet.

11

We have so far ignored the higher levels, though in the broad connotation of the term they include some of the most extensive upland plains in Britain, including the summit plains of the Welsh Mountains and the Grampians. If it be tentatively assumed that the platforms at and below 1,000 feet are of late-Tertiary marine origin, it is at these higher levels that we might hope to find representatives of one or both of the two sub-aerial peneplains of southeast England, the sub-Eocene and the Miocene surfaces, though it is far from certain that they would be separately recognisable. Both of them, and especially the first, might be expected to record appreciable warping. I think there can be little doubt that the summit plain of Wales includes representatives of one or both of these surfaces, Similarly, I note with satisfaction J. E. Richey's conclusion that, since the higher peaks of the Tertiary volcanic area of western Scotland attain to 2,000 or even 3,000 feet, the igneous rocks were involved in the great summit plateau of the Highlands.[1] It would evidently be most remarkable, if not quite unaccountable, if the great periods of Tertiary sub-aerial base-levelling, and notably that which followed the Alpine folding in the south, were not represented in our upper mountain levels. There are, I believe, two extreme mis-interpretations of the evidence of which we have to beware. On the one hand I must reject both of the alternative suggestions that these high surfaces are exhumed sub-Cretaceous or sub-Triassic planes. Therein lies the old-established heresy of virtually ignoring Tertiary time. On the other hand I view with some disquiet what I may term the nibbling at the high summit surfaces from below by the multiplication of supposed platforms on map evidence alone. We must follow where the facts lead ; but let us be sure that they are facts and not overlook the wide spacing of our surveyed contours at the higher levels. We may recall that Hollingworth's altimetric analysis, taken as a whole, shows no major frequency maximum between 1,000–1,070 and 2,000

[1] J. E. Richey, 'The Tertiary Volcanic Districts' *British Regional Geology* (*Mem. Geol. Surv.*, 1935), p. 110

feet. On the other hand we must note that H. Fleet [1] found surfaces in the Highlands at 2,300–3,000 feet, 1,400–2,000 feet, and 750–1,000 feet. Hollingworth finds some evidence of planation at 1,170, 1,340, and 1,600 feet in the Lake District, and McConnell [2] distinguishes surfaces at 1,700–1,850, 1,500–1,600, and 1,250–1,400 feet in the Howgill Fells. Similarly, Miss Goskar [3] subdivided the Welsh high plateau at 1,450–1,500 and 1,200–1,300 feet, while more recently E. H. Brown finds some evidence of planation at 1,200–1,300 and 1,450–1,500 feet in the Plynlimon area. Among the results of this kind which I regard as most trustworthy, having watched the progress of the work in my own department, are those obtained by Balchin on Exmoor, based upon accurate field work. He claims recognisable surfaces at 1,000–1,225 feet and from 1,250 to 1,600 feet, the last being what I should call an 'envelope surface' tangential to the culminating points.

In his last most valuable and suggestive paper on the Dartmoor area J. F. N. Green [4] found the highest evidence of an unwarped erosion surface, conceivably marine, at about 1,150 feet. He suggests that the higher summits of Dartmoor between 1,564 feet (Rippon Tor) and 2,039 feet (High Willhayes) closely fit a plane which slopes 30° south of east at 1 in 130. He further suggests that this is compounded of an eastern slope of 1 in 160 and a later superimposed tilt of 1 in 260. This I conceive to be the type of evidence which ought to be yielded by our higher upland summits. It would be a highly suspicious circumstance if we could find no evidence of warping at the higher levels. The multiplication, on fragile and ramshackle evidence, of supposed flats up to the level of the highest summits would confront us with features extremely difficult to explain, and if their reality were demonstrated they would throw doubt on the interpretation of the better marked record at lower levels. In default of cogent evidence to the contrary I therefore accept the conclusion of Trotter [5] that the surface of the High Pennines in the Alston block is essentially a sub-aerial peneplain and, with rather more

[1] H. Fleet, *Rep. Int. Geog. Cong.* Amsterdam (1938), p. 91
[2] R. B. McConnell, *Proc. Yorks. Geol. Soc.* **24** (1939), 152
[3] K. Goskar, *Proc. Swansea Sci. and Field Nat. Soc.* **1** (1935), 305
[4] J. F. N. Green, *Proc. Geol. Assoc.* **60** (1949), 105
[5] F. M. Trotter, *Proc. Yorks. Geol. Soc.* **21** (1929), 161

doubt, Versey's similar conclusion concerning the summit surface of the North York Moors.

12

Even if we leave aside the more debatable evidence from the higher levels, the facts reviewed offer a highly significant body of data which challenges interpretation in the wide over-lapping sector shared by geophysics, stratigraphical geology, and geomorphology. On the general and geophysical aspect of the question at issue I do not consider that it is yet possible to add usefully to the shrewd and balanced discussion published by Professor Baulig in 1935.[1] If we seek support for the general hypothesis of late-Tertiary eustatic shifts of sea level, it can hardly be denied that Britain affords it in very notable degree. Baulig emphasised the wide recurrence of his three dominant levels, at about 180, 280, and 380 metres throughout France and the Mediterranean area. Of these at least the two lower members have close parallels in Britain, and the third is present, if less widely known. But the British record is in some respects both more detailed and, as I believe, more accurately based than that of France, so that it might fairly be claimed that it yields independent evidence for the eustatic hypothesis. J. F. N. Green [2] recently quoted J. Ball's estimates of the successive Pliocene and early Pleistocene sea levels of the Mediterranean (Middle Pliocene 591 feet, Late Pliocene 505 and 423 feet, Early Pleistocene 338, 236, and 187 feet) and noted how closely these correspond with his own findings in Britain.

But I must forego here any temptation to pursue the primrose path into the seductive quagmire of long-distance correlation. There are, for the moment, sufficient British problems of correlation confronting us and we have by no means yet done sufficient field work to settle them. The cumulative force of the evidence in favour of eustatic movements is indeed very strong, but it cannot yet quite sustain a rigorous proof of the absence of warping, nor, on the other hand, does it suffice to locate or evaluate any such

[1] H. Baulig, *Inst. Brit. Geog.* Publication No. 3 (1935), and in 'Problèmes des Terrasses' *Union Géographique Internationale* (1948)
[2] J. F. N. Green *Proc. Geol. Assoc.* **60** (1949), 105

warping. It is clear enough that among the workers of the 'platform school' there are found representatives of both those widely spread types of mind which we may term the 'lumpers' and the 'splitters'. The two points of view need not be unalterably opposed, but it remains true that whether or not one finds apparent evidence of warping depends, often enough, on the number of distinct platforms which one is prepared to recognise. We do well to remember therefore, that if, as seems probable, we have evidence of one or more advances and regressions of the sea, the sea has stood at least once, and perhaps more than once, at every level within the total range of height affected, and a great multiplication of platforms is to be expected. Nevertheless, as Miller has emphasised, the final stages of retreat must necessarily have left the higher members of the series exposed to sub-aerial attack and a certain breaking down or semi-grading of the marine stairway must inevitably have followed. The resulting surface, however, cannot, in my view, be properly termed a peneplain save in a purely descriptive sense. It is in no sense comparable with that long-term product of sub-aerial wastage which we conceive as terminating a true cycle of erosion, but is simply a composite surface of marine planation modified by sub-aerial exposure.

A further problem of interpretation is raised by the question : Were the strand-flats cut during the advance or retreat of the sea, or during both ? Here we must not ignore the general evidence of stratigraphical geology. I have always rejected in my own mind the facile conclusion that as we climb the platform stairway we move steadily backwards through Tertiary times, so that if the lower platforms, up to say 700 feet are taken as Pliocene in age, those immediately above them are, *ipso facto*, Miocene. No facts at present before us really warrant this conclusion though it remains a possible hypothesis. But if we choose to entertain it we must recall that Europe as a whole gives evidence of distinct Miocene and Pliocene transgressions separated by a phase of regression. At its greatest extent, in Middle Miocene times, the Miocene sea was certainly not far from the British area, but I have no doubt that geologists will remain sceptical of its actual presence within Britain, unless palaeontological evidence comes to hand. In any case it seems an inescapable conclusion that, if any of our higher 'strand-flat' features were formed in Miocene times, both they and much ground at lower levels must have been bared to sub-aerial attack before any Pliocene transgression. For my own part I am content for the present to remain warily agnostic as to any recognisable Miocene phases and to conclude that, at least over much of western Britain, a Pliocene transgression, probably equivalent to the Calabrian stage of the Mediterranean, left evidence of planation up to a height of 1,000 or even 1,200 feet, and that the successive platforms at lower levels mark stages of retreat. I am not satisfied that we have yet devised sound criteria which might enable us to distinguish an 'ascending' from a 'descending' sequence of platforms, but the mutual relation of our platforms below 1,000 feet certainly suggests a descending sequence.

If this conclusion be accepted there is some suggestion of a differing upper limit of marine planation in the western and southeastern areas respectively. I can see no evidence at all of marine planation in southeast England above 700 feet, but many indications that the higher areas have been sub-aerially fashioned throughout. There is, no doubt, ground for difference of opinion here, for at best the evidence may be deemed indecisive. But while any doubt remains we must keep an open mind on the possible presence of warped surfaces. The general tenor of the British evidence as a whole admittedly points clearly to stability over wide areas with either uniform uplift or uniform regression of the sea. The local tilting or faulting so often invoked in older interpretations of British physiography is, in general, quite unaccceptable. If I may borrow from the parlance of the essentially geomorphological game of golf, it has often constituted a sort of tectonic 'niblick' employed in emergency to get the investigator out of 'bad lies'. But that is not to say that over wider areas we have yet attained a final demonstration of absence of warping, though I still hope and believe we may do so. It will require more than analysis of existing maps. Careful and objective field work alone can fill up the gaps in our evidence.

13

Whatever doubt remains concerning the detailed chronological interpretation of the evidence, there can be no doubt at all concerning the nature of the general physiographic picture revealed by our upland plains. They throw almost the only available light on the earlier stages of coastal development and much also upon drainage evolution.

It has long been an anomalous gap in British physiographic history that the stages linking the Tertiary coastlines with those of Pleistocene times and the present day are almost completely unknown. Jukes-Browne,[1] noting the evidence of the northward retreat of the Pliocene sea in East Anglia, assumed a general regression in Upper Pliocene times to the vicinity of the 100 fathoms contour line. This conclusion has been widely and uncritically accepted, but it is wholly erroneous. Apart altogether from the evidence of marine platforms, it is inconsistent with the known history of Pleistocene times. It would imply—and this conclusion has in fact been drawn—that all our coastlines and their Pleistocene predecessors were fundamentally 'coastlines of submergence', a general positive movement of the strand-line having over-ridden the fluctuations due to the varying volume of the ice-sheets. Of such movement there is neither any evidence nor even any plausible explanation. In fact, as we have seen, the Upper Pliocene coastline, successor to earlier and higher strands, was probably roughly parallel to the present coast and nowhere far from it. This conclusion is wholly consonant with the distribution of the earlier Pleistocene raised-beaches. From the maximum stage of the Pliocene transgression the general nature of British coastal evolution is clearly indicated by such work as that of Sparks in Sussex and Balchin on Exmoor. The sea withdrew uncovering in sequence a series of 'Ararats'. In Cornwall the 'Lands End' of the 400-foot stage was on the western side of the St Austell granite, leaving the Carnmenellis and Lands End granites as offshore islands ; at an earlier stage of higher sea level they may have been archipelagos closely comparable with the present Scilly Islands.

14

The related problems of drainage evolution are distinctly more complex. In some areas, e.g. the flanks of the London Basin at the 600-foot stage, the drainage direction on the contemporary

[1] A. J. Jukes-Browne, *The Building of the British Isles* (Stanford, 4th ed., 1922)

land was normal to the strand-line and emergence brought simple extension across the strand-flat of streams draining to the former coast. Even in this area there may be some doubt as to how far marine planation and the spreading of a thin sedimentary veneer completely obliterated the pre-existing topography. In some cases at least the extended rivers must have followed the line of their former courses fairly closely. But taking the south-eastern area as a whole, in the light of Linton's work in Wessex,[1] there is little doubt that we have evidence of an episode of 'wholesale' superimposition.

We have only to turn to Cornwall and Devon to perceive much more complicated and enigmatical drainage relations. Here is no simple case of coastward drainage 'extended' stage by stage over a descending series of strand-flats. In profile the rivers admittedly give the clearest evidence of an intermittent negative change of base-level. As shown by Dewey[2] the innumerable minor coastal gorges, well illustrated near Tintagel, mark the response to emergence of the '400-foot platform'. The Lydford gorge is similarly incised in the Bodmin Moor platform, while on the eastern side of Dartmoor the Becka Brook descends in a waterfall from the Bodmin Moor platform and traverses a deep gorge to join the river Bovey. In plan the relation of the rivers to the platforms is far less evident. The southward flowing trunk-streams (Exe, Tamar) rise close to the northern coast and appear to reflect a southerly tilt of the peninsula not shown by any of the platforms. This tilting may well date back to Oligocene or early Miocene times. If so we can only conclude that, as the later emergence proceeded, the rivers resumed their old south-ward flowing lines undeterred by the consequences of marine planation. If this was the sequence of events it affords independent physiographic evidence that the platforms are relatively narrow strand-flats and never extended far beyond their present outer edges. In other words, it would seem that the transverse depressions of the peninsula, such as that followed by the Tamar valley, must have been early established by earth movement, possibly faulting, and further developed by sub-aerial erosion before the Pliocene transgression. The Bovey Tracey-Petrochstow

---

[1] D. L. Linton, *Scot. Geog. Mag.* **48** (1932), 149
[2] H. Dewey, *Quart. Journ. Geol. Soc.* **72** (1916), 63

depression, which crosses the peninsula from side to side and may be of the nature of a rift valley, may well afford us a clue as to the nature and age of such features. It is pre-Upper Oligocene in date and, as Clement Reid surmised,[1] the associated faulting may well be contemporary with the Alpine folding of southeast England.

Even upon this hypothesis the rivers of the southwest still present many anomalies ; even head-water streams at comparatively high levels follow lines more nearly parallel than normal to former coastline features. Moreover there are many cases of diverted courses, simulating the effects of normal river-capture, but evidently related to the successive stages of the retreating strand-lines. An instance in which such diversion is imminent is seen near Bude. Around Bridge, some five miles west of Holsworthy, the Tamar is incised in a plateau with summits at 450 feet and with its valley floor at 300–350 feet. A mile or less to the west lies the degraded 400-foot cliff of the Bude embayment, and it is in course of sculpture by minor streams descending steeply to the coast. The westward diversion of the head of the Tamar at a point near Whitstone and Bridgerule Station may well ensue at no distant date. The Fowey today turns westward by a marked elbow north of Liskeard. A former southward course is marked by a wind-gap at 643 feet at Redgate. It turns sharply south again north of Lostwithiel, but a former continuation of the westward course below 500 feet is readily traceable. Barrow considered that this continued westwards to Newquay Bay [2] ; if so it was early diverted southwards by the Luxulyan stream and subsequently by its parallel neighbour draining south past Lostwithiel, A third diversion which caused a major modification of the morphology of the peninsula, was the reversal of the Taw, formerly tributary to the Exe, near Crediton, though the relation of this episode to the stages of retreat of the strand-line is yet to be worked out.

15

The light thus thrown on coastal evolution and river development suffices to show how vital to the general explanatory

[1] C. Reid, 'The Geology of the Country around Newton Abbot' *Mem. Geol. Surv.* (1913)   [2] H. Dewey, *Quart. Journ. Geol. Soc.*, **72** (1916), 63

description of British land-forms is the evidence afforded by the platform sequence. Unless or until geographers adopt the attitude of Gallio and decree that they care 'for none of these things' the general geographical justification of my theme is evident enough. But the sequence of land-form evolution I have reviewed has other and even more direct implications for geographical analysis, in its bearing on soil formation and land classification. If our successive upland plains represent surfaces exposed for different periods to atmospheric action, this must plainly be a factor in pedogenesis. I recently ventured to call the attention of pedologists to this factor [1] and by way of illustration cited the observations of Saint in the Barbadoes.[2] In an area where the parent material is coral limestone the soils grade from red above 700 feet, through intermediate stages, to black below 200 feet, and the facts were interpreted as marking stages of uplift of the island and the varying exposure to weathering of the successive emergent zones. Whether or not the physiographic hypothesis is, in this case, sound, the idea of a 'maturity sequence' is plainly implicit and must obviously be valid elsewhere. Recent years have seen some divergence in approach and emphasis between soil chemists and geologists in the study of soils. There has been a tendency to minimise the importance of the geological control of soil character and to call attention, quite rightly, to the numerous cases in which soil character is far from a simple and direct reflex of 'parent material'. There are, of course, many well-known reasons for this which I need not mention here, but among them is the question of the 'age of the site' in a much fuller and more fundamental sense than pedologists generally imply in using this phrase. The pedogenic processes, both geochemical and bio-chemical, are, within limits, cumulative in effect and functions of time. Is it not an inescapable conclusion that two surfaces, developed on the same parent material, of which one has been exposed twice, three times, or even ten times as long as the other, will show significantly differing soil profiles? We cannot claim *a priori* that this factor must be dominant, but merely that it exists and must be allowed for. It is not unlikely that, in general, bio-chemical equilibrium is attained more

---

[1] S. W. Wooldridge, *Journ. Soil. Sci.* **1** (1949), 31
[2] S. J. Saint, *Barbadoes Agric. Journ.* No. 3 (1934)

rapidly than geo-chemical in the soil profile, and may mask or even completely overpower the effects of differing duration of weathering. It is clear, moreover, that many of our upland surfaces have been substantially modified not only by glacial erosion and deposition but also by long-continued human action. But it is no answer to the point I am making to surmise, without further inquiry, that the effects would be trivial or vestigial. We should expect them to be most clearly in evidence as between sub-aerial, marine, and 'fossil' surfaces where such are juxtaposed. Southeast England [1] presents us with just this case ; the crest region of the Chalk, above the main Pliocene coastline, the Pliocene shelf itself, and the stripped sub-Eocene surface of the lower dip-slopes are found in close association. So far as 'solid geology' is concerned all these three areas are developed in Chalk, if we ignore the quite scanty relics of the former Pliocene cover. The geographer who is content to take his physical basis in crude and over-simple terms cannot but be aware that these so-called 'Chalk areas' differ widely, both in semi-natural plant cover and land utilisation. He is tempted to explain the differences solely in terms of the vagaries of human activity and to don a self-righteous halo for avoiding the heresy of determinism. But in fact it may readily be shown that the actual soil conditions differ widely on the three distinct surfaces. The sub-Pliocene and sub-Eocene surfaces are almost dramatically juxtaposed in east Kent with a hedgeless 'champion' landscape on the latter and a near approach to 'bocage' on the former. The contrast, however, is not merely in the soils of the residual flats, but in every detail of dissection and morphology. The contrast in the Miocene and sub-Pliocene surfaces is apparent both in Surrey and west Kent and also in the Chiltern Hills. The true clay-with-flints is largely, if not entirely, confined to the former surface.

Time will not avail me to multiply examples of such contrasts. I have observed them, in terms of soil profile, over the great expanses of the Weald Clay in Surrey and Sussex where the surface of the 200-foot platform reveals plain evidences of stronger

[1] S. W. Wooldridge and D. L. Linton, *Journ. Geomorph.* I (1938). 41, and in 'Structure, Surface and Drainage in Southeast England' *Inst. Brit. Geog.* Publication No. 10 (1939).

leaching than lesser and lower surfaces of more recent date. In this and other cases I believe that a detailed survey of soil-acidity would yield a pattern closely correlative with that of morphology. Nor can I doubt that maturity sequence expressed in soil will prove a valid factor in differentiating our upland moors. The several plant communities and their variants respond, of course, to the direct effects of form—of flat and slope and their immediate consequences in hydrology and micro-climate. But these same differences have been at work for even longer periods on soil and sub-soil and have left a legacy which we must disentangle from that of other factors.

16

Here then, if such be needed, is the direct 'economic' warranty for the concern of the geographer with the details of denudation-chronology, and I may finish where I began by stressing the proper geographical relevance of the subject I have sought to treat. The story of our upland-plains continued in that of valley-floor flats and terraces, is admittedly an 'evolutionary study' and indisputably a contribution to geological history. But without it there can be no 'explanatory description' of many British land-forms, nor will the geographer know sufficient of the nature and origin of the very land-surface itself to treat effectively of its successive human transformations. It may be an overstatement to say that all geographical knowledge, physical or social is, in the final analysis, historical knowledge, but it expresses a vital truth. I reject, in any case, the view that it is *only* as a means to a rational method of description of the earth's surface that geomorphology serves the geographer. We certainly are committed to study it as 'the field of organic activity and the abode of man', but we cannot do so without the fullest acquaintance with the manner of its fashioning. While, therefore, the 'platform story' has as much to contribute to geology as to geography I do not find it un-accountable that geographers have been actively interested in the problem I have surveyed. In any case I hope you will find it possible to bear with the attempt I have made as your President to maintain the link of our subject with geology and the related natural sciences to which it has always owed so much.

The Physiographic Evolution
of the London Basin

THE London Basin affords an admirable example of the evolution in time of a clearly defined natural region. The earlier stages of growth are of interest to the geographer only in so far as they contribute to this conception, but as such they are well worthy of attention.

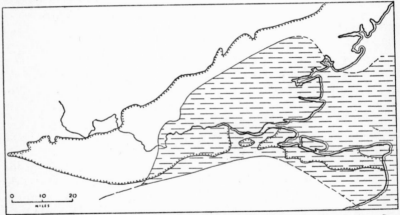

Fig. 4   Map showing geography of Thanet sands times in the London Basin.   Sea area shaded.   Edge of main Tertiary outcrop dotted

The most striking feature in the development of the area has been the constancy or frequent recurrence of major features of form throughout an immense period. Thus, there is definite evidence that in pre-Tertiary times a clearly marked synclinal basin existed in the area, drained by a stream analogous and ancestral to the present Thames. The surface over which this stream flowed was an extensive peneplain, preserved almost intact beneath the cover of Tertiary rocks and surviving also over considerable areas of the bounding Chalk Downs from which the Tertiary cover has been stripped.

The earlier stages of Tertiary development can be followed in some detail. The earliest Tertiary sea, in which the Thanet sands were deposited, occupied an area roughly co-extensive with the seaward end of the present Basin, though its axis lay south of the present middle-line. We see here evidence of down-warping of the old-established syncline and a resulting trans-gression of the sea from the east. A river of major size entered from the west (Fig. 4). In the immediately succeeding phase, in which the lower part of the Woolwich and Reading beds were formed, we note an advance of the sea in Surrey, Essex, and east

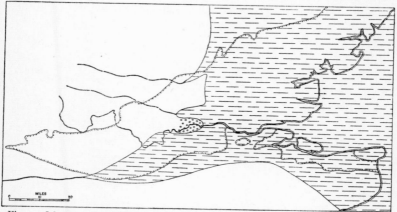

Fig. 5 Map showing the geography of early Woolwich and Reading beds times in the London Basin. Sea area shaded. Shingle foreland in circles

Hertfordshire, coupled with a retreat in south Hertfordshire and Middlesex. West of the coast-line shown on the map fluviatile sands and gravels accumulated, forming a broad sub-aerial delta-plain. The main river still entered from the west and the position of two other contributing streams can be fixed with some cer-tainty (Fig. 5). An interesting detail of the coast-line which has survived in recognisable form, though buried beneath the London Clay, is a shingle foreland projecting eastwards in the present neighbourhood of London—a veritable 'fossil Dungeness'. The upper portions of the Woolwich and Reading beds mark a further important change in geography. The sea retreated to south Essex and north Kent, where it occupied two gulfs, separated by a low emergent area over the region of the present

Medway mouth (Fig. 6). Herein we note the early establishment of the doubling or bifurcation of the main syncline, which is a feature of the present Basin and of the intervening 'Thames valley anticline', which also figures prominently in later times. The landward margins of the two gulfs were occupied by broad, brackish lagoons, while beyond, the sandy delta plains had passed to a condition of marsh, across which the rivers still flowed.

Though we cannot pursue the theme in detail, it is of interest to note that many of the physiographic features of the foregoing

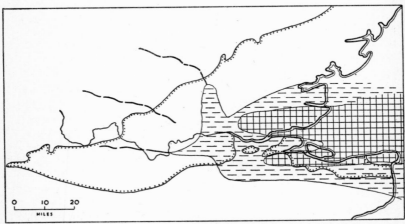

Fig. 6 Map showing the geography of later Woolwich and Reading beds times in the London Basin. Sea area shown by cross-shading, fringing lagoons by horizontal lines

stage were directly related to structural lines in the present Basin ; we see a picture of growing anticlines and synclines which, in their completed form, are important elements in present-day structure and relief. The Thames valley anticline has already been instanced ; a further striking example is the northward gulf-like extension of the 'Woolwich lagoon' (Fig. 6), which is a veritable ancestor of the present lower Lea valley. Both are related to a persistently unstable north–south 'ruck' in the Basin, to which further reference will be made.

A probable reason for the persistence of these structural features is that they are surface expressions of buried structures in the floor of older rocks beneath London. These rocks are undoubtedly much folded and faulted, like their analogues in the South Wales coalfield and the Ardennes. The newer rocks form a relatively thin skin over this buried nucleus, and it has wrinkled whenever the older mass has moved.

The next phase in the history of the Basin was that of the Blackheath beds—great masses of sand and shingle which mark a readvancing sea. These deposits, accumulated only over the site of the former lagoons and shallow sea-bottom, are hence

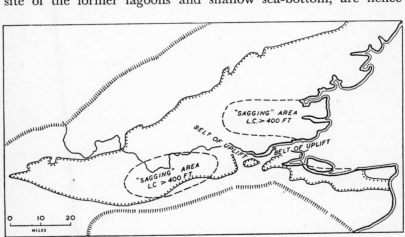

Fig. 7   Map showing position of sagging areas and belt of uplift deduced from the (original) thickness of the London Clay

not seen over the greater part of the Basin ; but the marine advance continued, carrying the sea over the emergent delta flats to the west and north and thus ushering in the London Clay phase.

This marks the greatest submergence of the area, during which the London Basin lost its individuality and was merged in the larger Anglo-Gallic Basin, extending eastwards to the Ardennes, southwards to Paris, westwards into Dorset, and northwards beyond the Chiltern Hills. However, if we note the distribution of those areas in which the original thickness of the London Clay was greatest—i.e. where the bottom was subsiding most rapidly, certain ancestral traits of form are found to persist (Fig. 7). These areas are respectively eastern Essex, north Surrey, and north

153

Kent (Sheppey). They are separated by two zones of relatively attenuated deposition, one following the lower Thames valley and the other crossing the Basin from northwest to southeast, passing through London itself. The first we readily identify as a further expression of movement along the Thames valley anticline. The latter is a symptom of a tendency, apparent even in pre-Tertiary times, for the Basin to 'break across the back', subsiding on either side of a central transverse ridge.

The last stages of the early Tertiary sedimentation saw the accumulation of the Bagshot Series—mainly a sand formation. Though preserved in force in the west of the Basin, they are confined to scattered outlines in the centre and east, and the tracing in detail of coast-lines, etc., is thus rendered impossible. The one significant fact we may gather is that the north–south Lea valley 'ruck' determined the position of the coast-line for part of the time, separating the marine Bagshot sands of Essex from their fluviatile or deltaic equivalents farther west.

This, very briefly, is the history of the accumulation of the ' solid' as distinct from the ' drift' materials out of which the landscape of the Basin has been carved, and upon which many of the geographically significant features of soil and scenery depend. It is vital to perceive that the succession of actual rocks varies considerably between the parts of the Basin, even though their age-names remain the same. The important features in considering the morphology and human geography of the region are :

(*a*) The Thanet sands disappear westwards near Guildford and nowhere appear on the northern outcrop, but in east Kent they are prominent, giving rise to land of high agricultural value.

(*b*) The Blackheath beds which dominate the physiography of southeast London have a restricted areal range, terminating abruptly westward along a line through Croydon, and being replaced by their sandy equivalents east of the present valley of the Medway.

(*c*) The London Clay thins steadily westward ; beyond Great Bedwyn, in Wiltshire, the Bagshot sands rest directly on the Reading beds. It is thus a much less important geographical element in the west than in the east of the Basin.

The uplift which excluded the early Tertiary sea from the region marked the beginning of the main formative stage in which the present structure was brought to completion. There is a 'lost interval' here, unrepresented by local deposits or recognisable physiographic traces. The precise date of the main 'tectonic storm' is unknown and need not concern us. We may take it as broadly contemporary with the Alpine mountain-building.

We may here abandon the chronological treatment for a moment in order to trace some of the direct effects of structure upon relief and drainage. Upon close scrutiny the structure of the region proves unexpectedly complex. There is a rectangular network of minor folds and faults, the main elements of which run with the grain or the strike of the country and are crossed at intervals by broader folds more widely spaced. In addition there is a series of gentle north–south monoclines or 'rucks' which play a part in shaping outcrops and relief features. The relation of the strike-folds to the drainage is evident in many places. The synclines generally form high ground, with anticlinal valleys between them—i.e. the relief is of the 'inverted' type. The effect of the monoclines is also striking. The lower Lea valley owes its striking asymmetry to its coincidence in position with one of these folds. It has been subject to 'uniclinal shifting', cutting steadily eastwards as well as downwards and leaving its successive gravel spreads on the western side. Other significant correlations of the monoclinal structures with outcrop and relief features will be apparent from the map (Fig. 8).

Following upon the main episode of folding there came a long interval of erosion, culminating in a peneplain—since uplifted to high levels. This feature, which is traceable all over southeast England and, indeed, over much of western Europe, is evidenced by the general accordance in the summits of the escarpments of the Weald and by much other evidence outside our immediate region. It represents the removal of a vast quantity of rock material and the complete obliteration of the relief produced by the folding movements.

The next legible incident in the history of the area was the advance of a late Tertiary (Pliocene) sea from the east into the synclinal tract. Its advent may be attributed to slight renewed downwarping along the old synclinal axis—and as such it is an

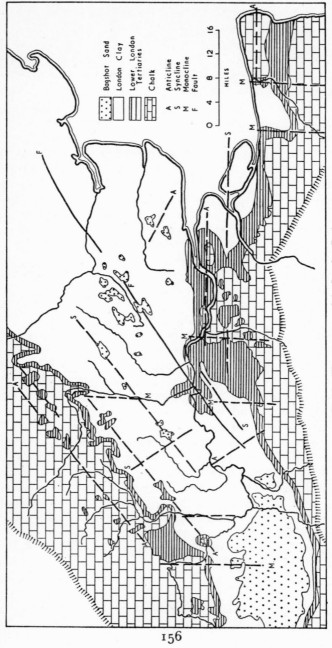

Fig. 8   The structure of the London Basin

exact repetition of early Tertiary history. The western connections of this sea, if any, are unknown and do not concern us here.[1] There was a general narrowing of the gulf or channel in the neighbourhood of London, due to a resumption of the old tendency to central transverse uplift, and as a result evidences of the old coast-line are preserved within the area bounded by the present Chalk escarpments, both in the Chilterns and the North Downs (Fig. 9). Eastwards and westwards of this the channel broadened and the exact positions of its margins are literally 'in the air'. The Pliocene sea has left its mark plainly on the landscape. The sand and shingle which it deposited

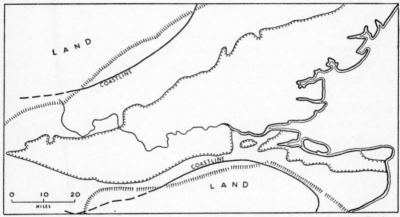

Fig. 9   Map showing the coastlines of the Pliocene channel in the London Basin

rest on a smooth abrasion platform which bevels the summit of the Chalk Downs—as in east Kent—or occurs as a shelf cut in the dip-slope, as in parts of Surrey, Buckinghamshire and Hertfordshire. The platform is merely the surface of the pre-existing peneplain itself trimmed by the waves of the sea. The only surviving portions of the peneplain itself within the confines of the Basin lie between the crest of the Chalk escarpment and the inner edge of the marine shelf ; a comparatively steep increase in the surface-gradient of the Downs occurs at the latter line.[2]

[1] cf. p. 97 and Fig. 1
[2] See profile sections of the North Downs and Chiltern areas in *Proc. Geol. Assoc.* **38** (1927), 55, 62, 84, 87

The ensuing chapters of the history of the Basin record essentially the uplift of the Pliocene marine platform to its present elevated position (500–600 ft. o.d.). The several stages of uplift and rejuvenation were separated by lengthy pauses, during which base-levelling became far advanced. Each such pause is marked by a platform or terrace. The countryside of London and the Home Counties may thus be described from the genetic standpoint as a series of maturely dissected plateaux, the formation of each having involved the partial destruction of its predecessors. Together they form a gigantic physiographic stairway, whose highest tread occurs on the bounding Chalk Downs and whose lower members are the familiar terraces of the Thames and its tributaries.

The first of the series of uplifts was slight, resulting merely in the emergence of the Pliocene sea-bottom and the extension of streams, formerly entering the sea, across it. These streams were flowing on a thin veneer of sand and shingle and paid little or no attention to the structure of the underlying rocks. This drainage system was directly, though not immediately, ancestral to the present system, and some of the anomalies of the latter can be explained only by the fact that it was originally superimposed from the Pliocene cover. Some of the features of the ancestral system can be traced by examining the provenance and distribution of the pebbles in the high-level gravels of north London. These deposits rest on a river-cut platform slightly lower than the marine Pliocene shelf and trenched within it. They show that the main stream then lay far north of the main synclinal axis ; its affluents from the Weald carried easily recognisable débris across the site of the present Thames valley to situations twenty miles to the north of London (Fig. 10).

A further and more pronounced uplift initiated a phase of down-cutting to the extent of about 200 feet, and led to the production of an extensive plain, diversified by relict hills. It must originally have been little above sea-level, but since has been lifted to 200–250 feet o.d., and may be called the 200-foot platform. Its dominance as a relief feature is apparent on any contoured map of the area, for though much dissected, many outliers of it occur ; its reality may further be amply demonstrated by planimetric methods, which show the abnormal preponderance of land at this height. It may be traced throughout the Thames valley, from the mouth upwards, and extends widely northwards beneath the glacial deposits of Essex and Hertfordshire. It is recognisable as a bench in the water-gaps of the northern Weald and southwards of these extends widely over the Weald Clay areas. The reconstruction of the drainage plan for this phase or during the preceding downcutting presents great difficulties, which are complicated by the glacial drift that covers much of the platform. The problem is, however, both interesting and important, and some treatment is necessary in an account of the evolution of the region. Of the suggested solutions we may note

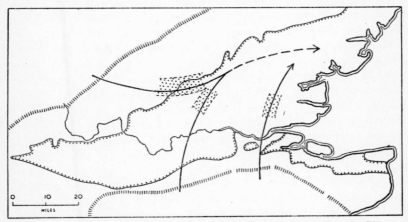

Fig. 10   Map showing the drainage lines of the London Basin after the Pliocene emergence.   The main tracts of high-level river gravels still preserved are shown in dots

first that of R. L. Sherlock.[1] He supposes that the Thames entered the Basin at Goring Gap, as at present, and hence followed a course northeastwards, immediately beyond the edge of the Tertiary outcrop as far as Ware, beyond which point it turned northwards again and passed *via* the Stort-Cam depression to the Wash. This view implies that the great belt of gravels mapped as glacial—stretching from Beaconsfield to Hertford—are in reality old Thames gravels. The assumption of the present

[1] For a full discussion of the glaciation of the London Basin and the successive shifts in the course of the Thames see S. W. Wooldridge and D. L. Linton, *Structure, Surface and Drainage in south-east England* (Philip, 1955).

drainage plan dates on this view from the main glaciation of the district, the ice advancing athwart the Old Thames course in the east and thus leading to the initiation of the lower Lea and lower Colne valleys as 'overflows'. Though we have seen that there is evidence that the Thames followed a northeasterly course in earlier times, the supposition that it did so at this much later date involves great difficulties. Prominent amongst these is the existence of two obvious wind-gaps in the London Clay escarpment south of Watford. These testify to a continuance of drainage *across* the line of the Vale of St Albans at a date approximate to that to which Sherlock's reconstruction refers. J. W. Gregory has presented somewhat different views, advocating the existence of an 'Ouse-Lea-Chelmer' river, independent of the lower Thames and following a southeasterly course from beyond Luton to the Blackwater estuary. H. L. Hawkins has made a further interesting suggestion, based upon the obvious analogy between the combined 'loop' of the Colne and Lea Valleys and the 'Henley loop'—a great incised meander, still occupied by the Thames. He has tentatively suggested that at a former date the Thames flowed, as it were, 'up the Colne and down the Lea', thus making a large-scale repetition of the 'Henley loop'. It is, of course, implicit that this course was followed at a level above that of the present valley floors. Even so, this attractive and stimulating hypothesis has to contend with numerous difficulties of detail, and if the evidence of the Watford wind-gaps is admitted, it certainly cannot be accepted. A cardinal principle of such reconstructions in the present area, and one all too frequently ignored, is that we may not invoke subsequent unequal uplift or warping to explain anomalies. That such warping has not occurred is apparent from the undisturbed condition of the older platforms. The lines of former streams must fall within existing physiographic depressions, and a reasonably even downstream gradient should still be demonstrable. Latterly, Mrs Ross has attacked the problem at issue in detail and arrived at conclusions with which the present writer is in close accord. Her reconstruction of the pre-glacial drainage is broadly similar to that of Professor Gregory, involving (*a*) a Kennet-Thames stream, flowing rather north of its present position and receiving amongst others two tributaries through the Watford gaps, and (*b*) a stream occupying the eastern end of the Vale of St Albans and flowing thence eastwards to the region of the Blackwater estuary. The alteration of this plan to the one at present existing is attributed to the interference of the ensuing glaciation in the east (as in Sherlock's hypothesis), coupled with normal river capture in the Watford area.

There is a certain amount of evidence of multiple glaciation in the London district, but we may pass over so debatable a topic in favour of a consideration of the features of what was certainly the main if not the only glaciation of the district, whose features are sufficiently evident. The ice-sheet deploying from the region

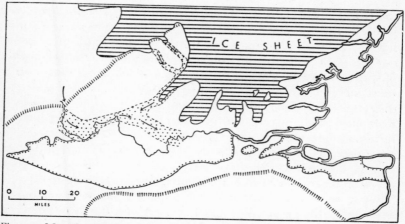

Fig. 11 Map showing the edge of the Ice Sheet in the London Basin. Outwash gravels are shown in dots and direction of transport is shown by arrows. Chalk escarpments and edge of main Tertiary outcrop are indicated

of the Wash overrode the chalk escarpment east of Hitchin and descended into the Basin, covering much of the 200-foot platform in Essex and Hertfordshire. Farther south and west, accommodating itself to the relief, it developed a rather complex edge, throwing tongues along the low ground intervening between the various residual hill masses. Its approximate margin is shown in Fig.11. In places, notably in east Essex and the Vale of St Albans, considerable tracts of outwash gravel accumulated in front of its advancing edge, and were subsequently covered in part by the boulder-clay. In the Vale of St Albans the gravels in transit southwestwards met and merged with the drift coming through

Goring Gap. One may regard the latter as 'long range' outwash, arriving by a longer route, or simply as the contemporary deposits of the Thames ; it is probably a distinction without much difference.

The direct effects of glaciation on the topography may be briefly mentioned. The onset of the ice east of Hitchin appears to have been largely responsible for pushing back the Chalk escarpment a distance of three or four miles as compared with its position in the Chiltern Hills. Within the Basin the drift-areas tend to a subdued relief, largely owing to the filling up of pre-existing hollows, but also owing to the amazingly flat surface of the boulder-clay sheet itself where undissected, as well exemplified in north Essex and the Vale of St Albans. The extreme flatness, however, tends to be modified by the close texture of drainage characteristic of the boulder-clay areas, and due to the occurrence of springs at various levels in the mass. Such areas thus tend to show incipient dissection, but in few cases indeed has maturity of dissection been reached or even approached. The edge of the drift sheet is characteristically featureless, being undiversified by terminal moraines or kame accumulations. The only possible exception to this statement is the curious gravel ridge of Tiptree Heath, between Chelmsford and Colchester. While this bears the semblance of a terminal moraine, most observers reject such a view, and many would favour a tectonic cause as a partial explanation of its position and form. Along the margins of the drift, particularly in Hertfordshire, several cases of diversion of drainage occur, of which perhaps the best is that of the river Mimram at Codicate, near Welwyn. Similar cases are known in the same region, but the details are not yet published.

After glaciation normal drainage conditions were resumed, the rivers occupying very nearly their present lines. There followed in normal sequence the deposition of the Boyn Hill (100 ft.) and Taplow terraces (50 ft.) of the Thames and its tributaries—both wide gravel-covered ledges. The Taplow terrace is the better preserved, having suffered less dissection, and it is covered by sheets of loam—either loess or mere inundation muds. Subsequently the narrow Flood Plain terrace was formed, and this was followed by the incision of a deep gorge along the lower Thames Valley (the Buried Channel). This long sequence of events is here only briefly summarised, since the features concerned are well known and have a large literature.[1] We recognise a continuance of those conditions of alternating uniform uplift and base-levelling which characterised the district from Pliocene times onward. The concluding phase in the physical history of the area was one of subsidence—most probably rise of sea-level—which resulted in the filling of the 'Buried Channel' with gravel almost to the brink of the Flood Plain terrace. During the later pre-historic and earlier historical phases of the region the estuary and its surroundings were still in a definitely submergent state and wide tracts were flooded at high-water. The present areas of marsh and alluvium are practically co-extensive with the former high-water sea-channels. As is well known, a considerable part of the silting up is of post-Roman date ; approximately seven feet of post-Roman alluvium was recorded in the Tilbury dock excavations.

Fig. 12 is a generalised drift-map of the Basin—indicating the very large areas wholly covered by the glacial and river drifts.

With the main facts of geology and morphology now before us, we may employ a method more purely geographical and attempt to sketch an outline of past environments of settlement in the area. Our real object is to reconstruct, as far as possible, the natural condition of the country during the earlier stages of its historic development. For this purpose a tentative division of the area into regions is proposed (Fig. 13). The regions reflect essentially unity of soil or vegetation, but though their ultimate basis is thus geological, their boundaries are not simple geological boundaries. We have been guided by considerations of the human responses evoked in finally delimiting them, particularly considerations of agriculture and population. The map as presented is a first attempt at subdivision, and thus inevitably a compromise. All that is claimed for the inter-regional boundaries is that they are truly 'mappable' lines corresponding to a visible change in the character of the country, and are based on long-continued field examination. It is, of course, evident that London itself has obliterated and absorbed parts of the original regions over which it extended, and that as a result factors once important have now vanished or are insignificant. For our present purposes, how-

---

[1] See S. W. Wooldridge and D. L. Linton, *op. cit.*

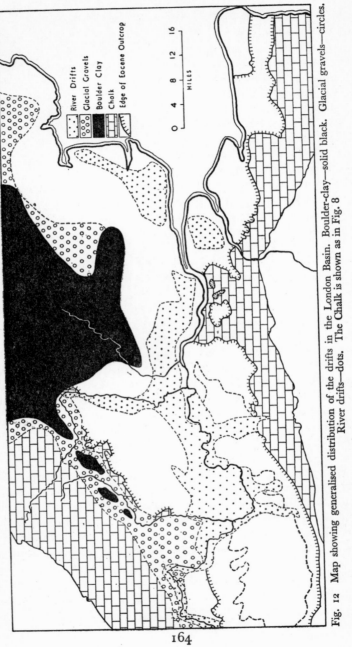

River Drifts
Glacial Gravels
Boulder Clay
Chalk
Edge of Eocene Outcrop

0   4   8   12   16
MILES

Fig. 12 Map showing generalised distribution of the drifts in the London Basin. Boulder-clay—solid black. Glacial gravels—circles.
River drifts—dots. The Chalk is shown as in Fig. 8

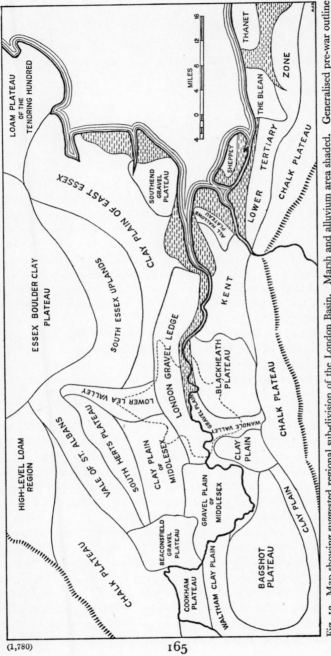

Fig. 13 Map showing suggested regional sub-division of the London Basin. Marsh and alluvium area shaded. Generalised pre-war outline of London shown in a dotted line

165

12

ever, we are ignoring the later growth of London, though its pre-war 'built-up' outline has been placed on the map for reference.

The bounding tracts of Chalk plateau or dip-slope show a considerable unity in their features. They are in a state of arrested dissection owing to the dropping of the water-table with recession of the escarpment. The soil of the plateau-surface is largely derived from heavy clay-with-flints, whose natural vegetation is 'dry-oakwood'. Locally the soil is lightened by the incoming of sheets of 'plateau-brickearth', and in the eastern Chilterns the dominance of this material justifies the separate recognition of a 'high-level loam region' which possesses rich and easily worked arable soils. Except where agriculture was thus favoured, the Chalk plateau areas were and still are heavily wooded ; to treat them as a whole as 'downland' is a deplorable mistake, suggesting that many historical writers have not visited the ground they describe. True downland country is confined to the face of the escarpments or their near vicinity and forms only a narrow strip.

The edge of the Tertiary outcrop is a definite regional boundary along the south of the Basin as far east as Farningham in Kent. Beyond this point the eastward expansion of the Thanet sands gives rise to an important soil-region favourable for cultivation. The sands—loamy or marly in character—extend up the chalk dip-slope as outliers, and also northwards to the Thames at Gravesend, and their wash masks and enriches a large part of the surrounding Chalk surface. We have accordingly grouped the whole area so constituted as the Kent Lower Tertiary zone. It is clearly bounded northwards against the marshes of the Thames-Medway, etc. ; its southward limit is fixed by considerations of surface-utilisation and marks approximately the limit between cultivation and mixed oakwood and pasture.

Prominent among the other regions south of the Thames we have the well individualised Bagshot plateau—a sand and gravel region of somewhat irregular topography. Its vegetation comprises wide tracts of the 'oak-birch-heath' association, with much 'subspontaneous pinewood' of recent growth. Its surface condition is much affected by the occurrence of 'iron-pan' (ferricrete), which forms below the soil by the capillary uprise of chalybeate waters and which holds up surface drainage, converting wide tracts into veritable marsh. It should be noted that this sand country is separated from the similar sandy terrain of the northwest Weald by a narrow clay plain and the even narrower Chalk ridge of the Hog's Back. These form the only breaks in a stretch of essentially similar sterile heathland which extends for twenty miles, measured from north to south.

The Blackheath plateau shows soils of a light character, and its vegetation type, where it survives undisturbed, resembles that of the Bagshot country. It should be noted, however, that considerable tracts of London Clay are included within it. A salient feature of the region is its steep cliff-like edge on the north, east, and south.

The minor clay and gravel regions south of the Thames do not require detailed comment here ; their features resemble those of the larger and similarly constituted tracts north of the river, which we now proceed to consider.

The 'London gravel ledge' is largely made up of the fifty-foot terrace of the Thames, but includes also gravels at a lower level. London north of the Thames was confined to this ledge until well on in the nineteenth century, as witness *inter alia* the line of railway termini which follows its back and marks the limit of pre-railway expansion. The original advantages of this region were dryness or firmness of surface, coupled with readily available water in springs and wells. In the light of its characters Trevelyan is certainly not justified in using the term 'water-logged' without qualification in describing the Thames valley of early times.[1] The extensions of the ledge east and west of London, over which building is now rapidly expanding, are extensively covered with loam or brickearth—one of the richest and most easily worked soils in the Basin. Accordingly both were regions of market gardening and fruit growing in recent times, and must have early found favour as plough-lands during the earlier settlements of the area. The natural vegetation, of which barely a trace remains, may be conceived as light mixed woodland. The 'gravel-plain of Middlesex,' also largely loam covered, lies in part of a lower elevation than the 'ledge' and shows similar brickearth cultivation. The Southend gravel plateau is again of identical type.

[1] G. M. Trevelyan, *History of England* (Longman's, 1926), Chs. 1, 2, 3

To the north of the Thames-side gravel regions is a strip of London Clay country, broken by the gravel-loam belt of the Lea valley. It is broad in Middlesex, narrow in west Essex, and broadens again considerably towards the coast. It was formerly covered with heavy forest, portions of which survive. It was and is subject to sudden and spasmodic flooding after heavy rain or melting of snow. Norden [1] remarked of the soil of the clay lands of Middlesex that 'after it hath tasted the Autumn showers (it) waxeth both dyrtie and deepe', and it is still true that normal winter conditions render unmetalled trackways almost impassable. On the other hand, during drought all surface supplies of water fail. These well-marked characteristics have rendered the London Clay tracts true barrier regions. The great importance of the Lea valley gravel strip, which affords a way through this difficult country, needs no emphasis. On the whole, the amount of true clay land in the Basin is relatively small ; its extent has often been over-estimated by writers unfamiliar with the area, who have ignored the wide coverings of drift.

North of the Middlesex clay plain is the plateau of south Hertfordshire. It is clearly bounded on the south by a sharp fall in level and its northern edge is the so-called Tertiary escarpment. It consists of dissevered remnants of the gravel-covered 400-foot platform, and though the slopes are true clay-land, there is sufficient 'gravel top' in the area to render its agriculture and settlement plan distinct. The villages are of plateau type throughout, depending originally on water in the thin gravel sheet. A certain amount of heathland still lingers on the plateau, but much of the lighter ground has been brought under cultivation. The south Hertfordshire plateau finds its continuation or analogue beyond the Lea in the region here styled the south Essex uplands. The high ground is less continuous than in Hertfordshire, and the summits often have a considerable sand capping (Bagshot sands) as well as a sprinkling of gravel. Moreover, the broad intervening saddles of low ground lodged ice-tongues during the glaciation and are covered with boulder-clay. There is some diversity of detail within the region, which includes the Epping ridges (clay) in the west and the gravel terrain of Danbury and Tiptree Heath in the east ; but its essential unity

---

[1] John Norden, *Speculum Britanniae* (3rd ed. 1723), pp. 11, 12

is undoubted, and it shows exactly the same conditions of settlement as the Hertfordshire plateau.

The Vale of St Albans is one of the most clearly marked of all the regions, and its contribution to the historic evolution of the area is also very evident. It owes its existence to the composite drift tongue which intervenes between the Tertiary edge and the true Chiltern plateau. Floored by gravel at its western end and largely by boulder-clay in the east, it provided from early times attractive conditions for clearing and cultivation, ready supplies of underground water, and a number of admirable riverside sites for settlement. Ware and Hertford at the eastern end match Watford and Rickmansworth at the western end in general situation and relation to London, while St Albans, literally 'a city set on a hill', has enjoyed from its focal situation, distinction as a centre of pre-Roman and Roman defence and administration, and now takes rank as the obvious capital of its small region. Westwards the gravels of the Vale of St Albans are continued in the Beaconsfield gravel plateau, though at a somewhat higher elevation. The plateau is in a youthful state of dissection, and its flat 'gravel tops' carry extensive light woodland and considerable stretches of heath. Cultivation is subordinate ; there are no associated loams, as in the case of the lower river gravels, and the soils are dry and acid.

Eastwards we have the Essex boulder-clay plateau, whose geographical features and place in history have been the subject of a recent study.[1] We may summarise its features by saying that it is in all respects an extension of East Anglia, and represents the nearest approach of the natural and traditional wheatlands to London. Its southern boundary as plotted coincides with no actual geological line, but marks rather the edge of the flat plateau and the limit of extensive arable land. The loam plateau of the Tendring Hundred forms a natural extension of the fertile boulder-clay lands, but is distinct in its physique and relations to the coast.

Finally, we may note the general expanse of marshland about the Thames-Medway mouth, and along the valleys. Each separate unit of the marshland tract can lay some claim to

---

[1] S. W. Wooldridge and D. J. Smetham, *Geog. Journ.* 78 (1931), 243; reprinted here as Essay No. 11.

separate recognition in its influence on the history of settlement, as can the fringe of little 'island' or peninsular uplands, Thanet, the Blean, Sheppey, and the All Hallows ridge (Hundred of Hoo).

Against such a 'mosaic' background of physiographic regions, the historical geography of the region must be viewed and interpreted. Certain of the regions stand out as more significant elements of the framework. The areas favouring early agriculture were the Essex boulder-clay plateau, the Kent Lower Tertiary belt, and the loam-covered tracts of the Thames valley. The first two converge significantly towards London from East Anglia and Kent respectively—a factor involved in the broader relations of its site and one which undoubtedly helped to control the strategy of the Roman campaigns and the course of Romanisation. The London Clay areas were heavily forested, dangerously impassable in winter, and effectively waterless in summer. They are practically destitute of evidences of Roman settlement, and a study of place-names suggests that they did not figure in Saxon settlement until a relatively late stage. In still later times their waterless state kept the population low, though recent expansion has engulfed them. Lastly, the wooded Chiltern plateau, the Beaconsfield plateau, the Bagshot country, and the heathland of the northwest Weald combine to form a strip of relatively difficult country, well supplied indeed with water in parts, but deficient in suitable arable soils. It forms a barrier across the Basin to the west of London, and one whose influence on historical development is well worthy of study. It is extremely poor in evidences of Roman settlement, and appears to have limited the eastward migration of the West Saxons to a considerable extent ; in its northern part the Chilterns, a people of supposed Celtic stock, survived. Its existence must be borne in mind in connection with the Conqueror's encircling march around London and with the delimitation of the Norman earldoms, to which it bears an obvious relation.

The Glacial Drifts of Essex and Hertfordshire

and their bearing upon the Agricultural and Historical Geography of the Region

THE glacial drifts of the English Plain already possess a very extensive literature on the geological side which concerns more particularly their constitution, succession, and origin. Less has been written on the subject of the control which they exercise on the geographical features of the country. It is recognised that East Anglia and parts of Lincolnshire and Yorkshire, as well as extensive tracts west of the Pennines, are almost entirely drift-covered, and in these regions the influence of the drift upon topography, soil, surface utilisation, and water supply is sufficiently evident. It can be fully appreciated, however, only by contrasting these areas with neighbouring lowlands, unglaciated or at least drift-free. When this is done it is found that the margin of the drift-country almost always forms a natural geographical boundary, not perhaps a hard line but a narrow zone of transition. Good examples of this are seen in west Norfolk where the drift-sheet terminates against the gently rising dip-slope of the East Anglian Heights ; or again, in even more striking fashion, near Melton Mowbray, where the Lias of the Vale of Belvoir lies at the foot of the escarpment-like edge of the boulder-clay plateau near the line of the Fosse Way.

In the present communication we have sought to trace the similar contrast observable along the southern margin of the great drift-sheet of eastern England, particularly in Hertfordshire and Essex. So far as we are aware no extended consideration has been given to the geography of this line, which is undoubtedly the most clearly marked natural regional boundary in the northern part of the London Basin, separating northern Essex from southern Essex, and eastern Hertfordshire from western Hertfordshire.

The extension of the glacial drift southwestward along the Vale of St Albans carries the characteristic 'drift' geography along a well-defined belt into western Hertfordshire, and gravels, closely associated both in character and age with the glacial drifts, occur in force along the northern side of the Thames valley between the river Colne and Goring Gap.

We need not pause here to consider in any detail the mode of emplacement of the drifts—a topic pertaining to pure geology, and one, moreover, concerning which considerable doubt still exists. It may be well to state, however, that in the following pages the glacial drift is treated as the product of land ice. On this view the boulder-clay must be regarded as 'moraine profonde' formed beneath the ice, or, more probably as we think, following Lamplugh,[1] as the product of the slow decay of ice densely charged with detritus. Lamplugh regarded the shallow sea basins beyond our coasts as the areas of greatest ice accumulation and erosion, while 'the thickest sheets of lowland drift were laid down where the ice . . . thinned off towards its periphery'. With this view we concur. It carries with it the corollary that the boulder-clay is not co-extensive with the former ice-sheet, having accumulated in a broad marginal belt round the centres of ice dispersal The explanation of the outwash gravels presents no difficulty ; it is worthy of note that the amount of such outwash is relatively small in the district under notice.

East of Hitchin the chalky boulder-clay ice-sheet deploying fanwise from the region of the Wash overrode the Chalk escarpment, and no doubt it descended the Chalk dip-slope normally in the direction of greatest slope. Within the Tertiary country, however, it entered a region of irregular relief in which relict hills diversified a surface already far advanced towards base-levelling. Here it split up into various tongues whose distribution and motion were controlled by the form of the ground. In short, it would seem that the Tertiary uplands ranging westward from Danbury through Brentwood and across the Lea into Hertfordshire acted as a vast fender holding the mass of the ice off from the region of the present Thames valley, but allowing narrow ice-lobes, which must have had the dimensions and appearance of true glaciers, to push through the gaps between the several

[1] *Quart. Journ. Geol. Soc.* **76** (1920), 70ff.

masses of high ground. Thus, in the west, the ice-sheet split on the northeastern angle of the south Hertfordshire plateau, sending one branch along the Vale of St Albans and another down the lower Lea valley. In Essex, the Epping Forest ridge separated the Lea valley lobe from that of the Roding valley. Farther east a long ice-lobe projected southwards to Upminster between the high ground of Havering and Brentwood.

Another lobe occupied the low ground between Brentwood and Billericay. These several lobes in western Essex are disposed in a radial manner about a centre lying in the region of the upper Stort and Roding valleys, and it would seem that the check to movement imposed by the barrier of high ground resulted in some measure of congestion in this depression, rendering it a separate subsidiary centre of ice dispersal. In eastern Essex, where the ice-sheet advanced into low flat ground, the margin is not lobate but follows a simple course from Chelmsford eastwards.[1] The boulder-clay of the main mass and of the several lobes and tongues has been subject to a relatively small degree of post-glacial dissection. Associated outwash gravels appear beneath it in valleys trenched through the sheet, and locally extend far beyond its boundary. Their distribution is treated in more detail in the sequel. The characters of the drift-margin may best be treated under the following three regional headings:

A Eastern Hertfordshire, where the drift country gives place to the normal Chalk plateau.
B The Vale of St Albans, where the drift extends south-westwards, with Chalk on its northern and London Clay on its southern side.
C Southern Essex, where the edge of the drift lies on London Clay throughout.

We may begin our survey of the drift boundary at Hitchin where it crosses the Chalk escarpment. Fig. 16 illustrates the striking change in the position and morphology of the escarpment which takes place here. Westwards for a distance of nearly twenty miles it is a bold feature constant in strike. Its mor-

---

[1] For a detailed account of the glaciation of the district see S. W. Wooldridge, *Essex Naturalist* **21** (1927), 257

phology is complex in detail since it is breached by several wind-gaps, as well as by the marked salient about the Lea headwaters near Luton, but the range is essentially continuous. The summit level falls steadily from 845 feet beyond the Buckinghamshire border to 602 feet at Telegraph Hill above Hexton.

The Hitchin Gap is six miles wide at its mouth near Hitchin, and on its eastern side the Chalk escarpment exhibits a complete transformation in form and relations. Its summit-level is much lower, rarely exceeding 500 feet, and it is capped by boulder-clay to the summit. It is set back some four miles southeastward of its former line of strike, and from its foot there stretches a

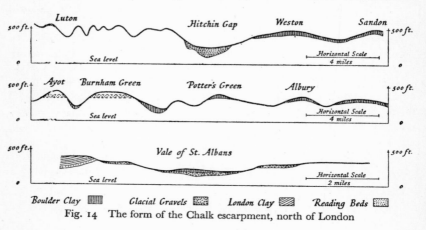

Fig. 14 The form of the Chalk escarpment, north of London

broad low-lying outcrop of Lower Chalk (Chalk Marl) which forms part of the Vale of Baldock. Its outline is characteristically subdued as compared with that of the escarpment to the west. This is clearly the area in which the change from the Chiltern Hills to the East Anglian Heights may most appropriately be deemed to take place. The change in the character of the escarpment is to be attributed largely, if not entirely, to the onset of the ice east of the Hitchin Gap, whereby the face of the scarp was pushed back, and the Chalk Marl plain at its foot formed. The latter is essentially drift-free, and is to be regarded as complementary to the boulder-clay sheet to the south, in the sense that its formation provided much of the material deposited beyond the escarpment crest as the chalky boulder-clay.

The map (Fig. 15) illustrates the distribution of the glacial deposits (boulder-clay and glacial gravels grouped together) on the Chalk dip-slope. East of the Hitchin Gap the plateau surface is covered by boulder-clay, though several wide valleys expose the underlying Chalk, and a few patches of clay-with-flints appear in the west. West of the gap the higher parts of the plateau are uniformly covered with clay-with-flints and plateau brickearth. The floor of the gap is filled to a great depth with glacial gravels. The actual western edge of the drift runs south-eastwards from near Hitchin past Stevenage, and finally turns through a right angle into the Vale of St Albans. The difference in soil character between the plateau areas east and west of the gap is in itself sufficient to determine appreciable differences in geographical character, but the contrast between the two regions does not reside only in soil.

The sections, taken transverse to the gap and its continuing valley (Fig. 14), indicate how plainly the glacial limit is marked in the morphology. The ridge running from Offley through Preston to Knebworth and beyond presents a steep face to the east, and rises to a greater height than the boulder-clay plateau at Weston on the opposite side of the gap. It is a most marked physiographic line, conspicuous both on the map and on the ground, and may be regarded as marking the easterly limit of the typical Chiltern plateau. To the west the plateau rises to greater heights than are reached to the east, and the valleys are straighter, more closely spaced, and steeper sided. The general effect is to produce a topography markedly more accidented in the western or extra-glacial region. The subdued contour of the eastern plateau derives only partially from the smothering or levelling effect of the boulder-clay sheet. Its different morphology doubtless reflects in part the operation of glacial scouring and planation. Farther to the south the contrast between the two sides of the Beane valley is emphasised by the occurrence of a line of conspicuous Eocene outliers on the west, at St Albans, Ayot, and Harmer Green. The line of outliers certainly persists for some little distance to the east, but it soon fails, and the outliers which do occur are partly obliterated by the boulder-clay mantle.

An even more important difference between the two regions appears in their hydrography. All the valleys in the east contain

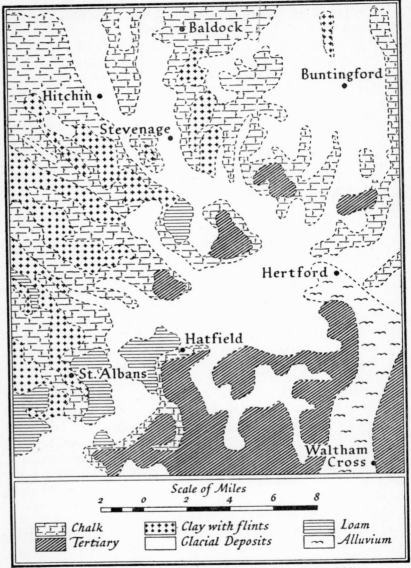

Fig. 15 Distribution of glacial deposits in eastern Hertfordshire

flowing streams, but on the west, while there is a complex system of valleys, they are all dry, with the exception of those of the main consequent streams. The reasons for this hydrographical difference are not entirely clear, for there does not seem to be any systematic difference in the position of the water-table. On the other hand, the boulder-clay is an effectively porous stratum over wide areas, emitting springs at its base and at other levels, and these, no doubt, serve to maintain a surface flow above the point of emergence of the purely Chalk springs. But whatever the cause of the difference, its effect on the habitability and early settlement of the eastern region is sufficiently plain.

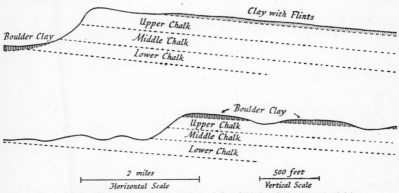

Fig. 16 Profiles across the margins of the drift-lands in Hertfordshire

Passing now to the Vale of St Albans, we find that the drift forms a parallel-sided strip about five miles wide. The boulder-clay is confined to the lower parts of the floor of the vale, extending as a broad dissected tongue as far west as Aldenham. The subjacent gravels are thick and continuous and extend beyond the boulder-clay tongue to Watford and westwards. They are disposed in two terrace-like spreads, one flooring the vale and the the other, more dissected, forming a step at a higher level, better preserved on the northern side (Fig. 14). The boundaries of the drift on either side of the vale are distinctly marked. On the south rises the so-called Tertiary escarpment, the edge of the dissected clay plateau of south Hertfordshire. The latter region is drift-free, if we exclude some thin patches of high-level gravels of pre-glacial date which have little influence on soil or surface-features. The boundary on the north side is distinctly more remarkable though no more striking in its geographical expression. Here the drift rises to an elevation of about 400 feet and terminates along a very nearly straight line which runs south-westward from near Welwyn, passing slightly north of St Albans and thence onwards past Chorley Wood to Beaconsfield, finally emerging on the northern slope of the Thames valley near Marlow. Throughout this long range the boundary maintains a strikingly uniform elevation, and it corresponds with a soil change of the most drastic character. Beyond, on the Chiltern dip-slope, nothing but clay-with-flints and kindred argillaceous materials occur on the plateau surface, but as soon as the drift-line is crossed we enter a region of deep sandy and loamy soils, which, while locally stony and sterile, have proved capable of yielding under treatment excellent farming land.

Finally, we may note the curiously symmetrical arrangement of the drainage in the vale. This is drained by the subsequent portions of the Colne and the Lea, flowing westwards and eastwards respectively. Each gathers a share of the consequent drainage of the Chiltern dip-slope, the Chess, the Gade, and the Ver entering the Colne ; while the upper Lea, the Mimram, the Beane, and the Rib take the eastward route. This simple arrangement has exercised its due control on the settlement of the area.

In Essex the topographic expression of the drift boundary shows considerable variations from that noted in Hertfordshire. In the latter area the drift generally abuts against rising ground ; but in Essex this is rarely the case. The map (Fig. 17) presents a simplified version of the geology of the region, and enables us to distinguish the varying character of the drift boundary. In the southwest of the county the boulder-clay rests directly on the London Clay, associated gravels being absent or subordinate. It lies for the most part at relatively high elevations, capping the ridges, and the sheet is much dissected owing to the proximity of the lower Lea and Roding valleys of which the various phases of rejuvenation have affected the whole area.

East of Brentwood the edge is still very irregular and gravels again are absent. The boulder-clay here rests in the broad cols between the outliers of Bagshot sand. Between Chelmsford and

Colchester, however, a marked change takes place in the general geological and physiographic setting, for large masses of gravel appear from beneath the boulder-clay and occupy extensive tracts in front of its edge. The hills at Danbury, continued beyond the Blackwater in the curious ridge of Tiptree Heath, follow the axis of this gravel terrain. This is the only part of the county in which the limit of the drift is marked by any clear topographic feature. It has been suggested [1] that the Tiptree

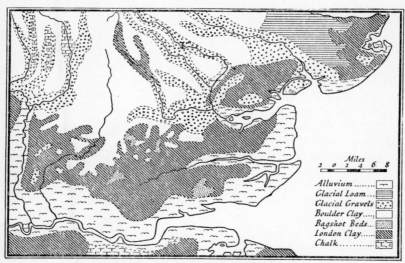

Fig. 17   Geology of Essex (simplified)

Ridge is morainic in character, but though its situation and relations lend some colour to this hypothesis more investigation is required before it can be accepted. Finally, in the Tendring Hundred, the area between the Colne valley and the coast, the boulder-clay edge retreats northwards into Suffolk and the place of the frontal gravels is taken by an extensive sheet of loam, which caps the low plateau of London Clay.

Certain variations are to be observed also in the character of the drift country itself in central and northern Essex. Between the Stort and Chelmer valleys the boulder-clay sheet is not broken through by the valleys, and a characteristic uniformity of surface

[1] F. W. Harmer, *Geology in the Field*, vol. I (Stanford, 1908), p. 122

results. It rests on London Clay without intervening gravels, and the twin head valleys of the Chelmer give rise to an enclave of true London Clay country in the centre of the drift plateau. To the east the parallel valleys of the Blackwater, Colne, etc., reveal the northward continuation of the frontal gravel sheet, the boulder-clay being confined to the summits of the broad interfluve ridges. The presence of extensive gravel tracts considerably modifies the character of the country. In particular the presence of these porous beds reduces the closeness of texture of the drainage system as compared with that of the boulder-clay area to the west.

A driftless London Clay lowland intervenes between the edge of the glacial drifts and the broad belt of the Thames alluvia which borders the northern side of the river. This lowland is narrow and irregular in the west but broadens eastwards into a curious flat-floored trough which trends northeastwards across the line of the Crouch valley to the Blackwater estuary. It is continued beyond the latter in the area around the Tolleshunts, though this outlying portion of the London Clay terrain has significant characters of its own. It is of great geographical importance that the drift boundary approaches the coast so nearly in eastern Essex and that the navigable estuaries of the Blackwater and the Colne give natural and direct access to it.

One very important physical character of the drift lands both in Essex and Hertfordshire is the readiness with which they yield a supply of water from shallow depths. The early settlement of the English Plain depended upon the availability of a water supply as much as on any factor. Excluding the regions of the scarp-foot springs there are few areas so favoured in this regard as those of the glacial drifts. Shallow wells yielded adequate supplies of water throughout northern Essex and eastern Hertfordshire for hundreds of years, until, in fact, within recent times it became necessary to guard against contamination and secure a generally higher purity. This water was obtainable both from the gravels and from the boulder-clay itself, which in virtue of its heterogeneous constitution is often water-bearing. The presence of a boulder-clay capping protects the gravel waters from contamination in many cases. Even where the base rock is Chalk a separate water-table is generally maintained in the gravels owing to the check in the downward percolation that takes place at the gravel-chalk junction. In contrast to these conditions we may note that over the Chiltern plateau the water-table in the Chalk is rarely less than 200 feet below the plateau surface and generally more than 100 feet below the floor of the smaller valleys. Again, the London Clay outcrops were effectively waterless areas, so far as underground supplies were concerned, until the era of deep wells.

In the foregoing paragraphs the physical contrast between the drift lands of Essex and Hertfordshire and the tracts which border them has been outlined. In pursuing this contrast into the sphere of human geography, we must first emphasise the fact that soil differences have been the most important factor in the geographic development, the hydrographic and mor-phological factors filling an important though secondary rôle. The essential fact concerning the drift lands under consideration is the ease and profit with which they have been cultivated, and it is therefore desirable to devote some attention to their soil features.

The geological map distinguishes three categories within the 'glacial' drifts : boulder-clay, glacial gravels, and loam or brickearth. This simple threefold grouping fails adequately to indicate on the one hand the wide variability of the resulting soils and on the other their underlying unity of character. Here we find illustrated at once the rather crude and ineffective nomenclature of sedimentary petrography and the prime need of soil maps for geographical work. Both aspects of the case are at present receiving attention, but it will be many years before the leeway in these branches of knowledge is appreciably made good. Meanwhile it is beyond the compass of individual workers to make soil surveys of large regions, a task which needs co-operative effort. We propose in this case to submit certain general facts and conclusions based upon a survey of the whole area in outline, together with the detailed examination of a selected tract round Welwyn in Hertfordshire where the several types of glacial drift together with the Chalk, clay-with-flints, and Eocene beds are exposed in close juxtaposition.

The boulder-clay is the most extensive of the glacial drifts of the region. Its general characters are well known and need not be described, but several important points concerning its variations demand attention. The term 'clay' is frequently all but a misnomer ; 'stony loam' would more nearly describe it over wide areas. Moreover the removal of the finer particles by long-continued rainwash on gentle slopes concentrates the coarser constituents and gives rise to a soil consisting of stones in a silty base not unlike the soil derived from the gravels. Inter-stratified beds of sand and gravel ensure further departures from the true clay type. Even more important is the fact that the character of the boulder-clay inevitably depends in part upon the immediately subjacent rocks. In all cases there is a far-travelled element in its composition, homogeneous, well mixed, and uniform over wide areas, and a local element whose nature and amount varies widely from place to place. Thus it is a fact familiar to geologists that the typical chalky boulder-clay of northern Essex and southern Suffolk gives place beyond Ipswich and Bury St Edmunds to a 'chalky Jurassic' boulder-clay in which débris from the Jurassic clays becomes conspicuous. In the area under notice an equally important variation is seen, but one to which less attention has been given. The ice was here moving transversely to the rock outcrops, and the boulder-clay sheet lies across the boundary of the Chalk and the Eocene beds, largely concealing it. The concealed boundary runs from Hertford *via* Bishops Stortford and Thaxted to Sudbury. Thus, in eastern Hertfordshire the main mass of the boulder-clay rests on the Chalk except between the valleys of the Stort and the Ash. In western Essex it rests upon the London Clay. As a con-sequence in Hertfordshire the boulder-clay is dominantly chalky, while containing a sufficient admixture of far-travelled elements to yield a soil richer than that derived from the Chalk itself. The dominantly chalky character persists for some distance across the Eocene outcrop, but increasing quantities of London Clay are present in the mass involving a marked change of character. Hicks [1] observed many years ago at Hendon and Finchley that the boulder-clay, though containing an appreciable chalky basis, was locally little more than reconstructed London Clay, large masses of the latter having been bodily incorporated by the ice. Thus it is that the southern margin of the boulder-clay tracts gives a colder and wetter soil less suitable for cultivation than the true chalky boulder-clay, though superior to the

---

[1] *Quart. Journ. Geol. Soc.* **47** (1891), 575

London Clay itself. In other words, there is a natural transitional zone which blurs but does not destroy the sharpness of the geographical contrast. The true chalky boulder-clay is today largely under cultivation, while those portions contaminated with Eocene material (chiefly London Clay) have reverted in part to pasture. The reality of the distinction is well brought out by a comparison of the country round Buntingford and the Pelhams with that round Hunsdon and Widford. It is further illustrated by the contrast between the boulder-clay at Harlow, still highly chalky, and that which covers the Epping ridges only five or six miles to the south. In the latter area and again west of the Lea above Waltham and Cheshunt the boulder-clay country is marked off by no salient characteristics from the surrounding tracts of bare London Clay. Recent sections on the main road north of Epping have exhibited the boulder-clay highly charged with London Clay and relatively poor in chalk.

Before leaving the subject of the boulder-clay soils we may note that east of the Chelmer valley and again in the Vale of St Albans the boulder-clay rests on its own proper lower member, the outwash gravels, disposed in front of the ice and subsequently overridden. Here it is to a large extent insulated from the local base rock, local variations in character are subdued, and its 'geographical contact' with the London Clay is correspondingly abrupt.

Passing now to the glacial gravels we have to deal with another extremely variable accumulation whose accepted name not infrequently belies it. Taken as a whole the gravels tend to be loamy, containing a large proportion of interstitial silt and clay as well as definite interstratified clay beds. The resulting soil is rarely so dry and acid as one might expect and, as noted above, often differs but little from the boulder-clay soils. Locally, as around Hitchin and west of Hertford, sand predominates over gravel, but even here the soil is loamy. Heathland is subordinate throughout the area and there is no approach to the 'breckland' type. It may be noted that locally the glacial sands and gravels are highly calcareous, though of course they readily suffer surface decalcification. On the whole there is good justification for grouping the gravels with the boulder-clay as yielding broadly similar soils. The gravel soils are generally less favourable to arable cultivation than the best chalky boulder-clay soils, though locally at least they are superior to the 'Eocene' boulder-clay. For instance, in the Symonshyde area between Sandridge and Welwyn the boulder-clay is left under pasture and woodland while the gravel soils are ploughed. It is further important to notice that the place of the gravel soils in the agriculture of the region has changed during the last century. Arthur Young, Secretary of the Board of Agriculture, in reporting on land farmed by himself at North Mimms in 1804, spoke very unfavourably of the gravel soils of the Vale of St Albans. He characterised this district as 'the most unfertile that we find in the south of England' and spoke of the wetness or 'spewiness' of the soil due to the many springs. He further noted that it yielded to manuring when drained but that the benefit was soon lost. The change in the value of the soils followed from the fact that they were warm and light and that their natural infertility could be made good at small cost with London stable manure. This character has more than outweighed their lack of inherent richness and, in the words of Sir Alfred Daniel Hall,[1] they have proved that they can 'rapidly convert expenditure on manures into crops commanding a good price'. In the Vale of St Albans the ready availability of water-borne manures from London as well as the proximity of the London market undoubtedly favoured the cultivation of the gravel tracts. The gravel areas of eastern Essex are in a less favoured situation and have never been cultivated to the same extent. Today they include some of the few large stretches of heathland in the whole area (e.g. Danbury Hill, Tiptree Heath).

The third category of glacial drift shown on the published maps is designated 'Brickearth (Loam) and Laminated Clay'. Its origin, whether through the agency of water or wind, has been deemed somewhat problematical. There is no doubt, however, that much of it is a true wind-blown loess, though the laminated variant has evidently been deposited in standing water, and various small patches on the boulder-clay surface may represent 'kettleholes' filled with rainwash, the complement of the static washing process. The representation of this class of material on the geological maps is far from satisfactory, though

[1] Sir Alfred Daniel Hall, *Victoria County History of Hertfordshire*, vol. II (Constable, 1900), p. 135

one can feel every sympathy with the surveyor in dealing with
so vagrant a class of drift. The largest mapped expanse occupies
the western part of the Tendring Hundred near Colchester.
Small areas are indicated at various places on the Essex boulder-
clay sheet near Witham, Finchingfield, and elsewhere. Of more
importance are the large tracts shown on the quarter-inch Index
Map (Sheet 16) on the floor of the Vale of St Albans, west of
Hatfield. Though not separately distinguished on the new one-
inch Drift Sheet (No. 239) the reality of these spreads is not in
question. The thickness is here about four feet and the deposit
rests on boulder-clay. Finally we must note the occurrence of
extensive tracts of a similar loam on the chalk plateau round
Caddington and Pepperstock near Luton. It is here that the
evidence for aeolian origin is strongest, for the deposits are at an
elevation of 500–600 feet above the sea and yield palaeolithic
implements which date them as younger than the neighbouring
deeply cut valleys. In the light of these facts Worthington
Smith's suggestion [1] that they are lacustrine must in any case
be abandoned. Identical deposits range across the Chiltern
plateau beyond the confines of Hertfordshire into Buckinghamshire. Lithologically and geographically they differ in no important respect from the valley brickearths of the Thames, Lea,
etc., for which an aeolian origin may also plausibly be argued.
Whatever their origin, however, these loams provide an ideal
soil for arable farming, and their influence on the economy of
the countryside has been marked. The contrast which the
upland loams present to the stiff and intractable clay-with-flints
associated with them is particularly notable, for the latter,
though a good arable soil, tends to be cold and late and is distinctly more difficult to cultivate. The well-known map with
which Vidal de la Blache illustrates his *Tableau de la Géographie
de la France* [2] sufficiently emphasises the importance of the
easily cleared and readily cultivated 'loess' and 'plateau limon'
soil. The existence of such loams is indicated on his map within
the East Anglian region, but we are here suggesting that the
loess belt has a more extended westerly range and that it should
be reckoned as an important factor in agricultural economy and

[1] W. Smith, *Man, the Primeval Savage* (Stanford, 1894)
[2] See note on p. 226

historical geography at least as far west as the eastern part of the Chiltern plateau.

Before leaving the subject of soils a word may be said regarding the clay-with-flints, which in its true and typical form is a simple dissolution residue of the Chalk. It yields a soil of great inherent fertility but difficult to work unless it be chalked. It was formerly the custom to obtain chalk from small 'bell-pits' sunk in each field and to spread it liberally on the land. Uncultivated land

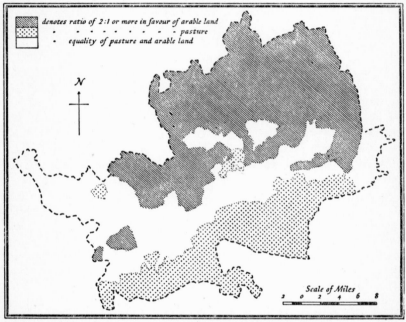

Fig. 18   Ratio of pasture and arable land in Hertfordshire

on the Rothamstead Estate 'proved to retain only 0.1 per cent of calcium carbonate, while cultivated land benefiting from former chalking had as much as 5 per cent.[1] The soil is naturally cold and badly drained ; chalking by coagulating the clay renders it warmer and earlier and facilitates drainage. Without it cultivation would have been virtually impossible in many areas. The practice of chalking has been largely discontinued

[1] Sir Alfred Daniel Hall, *op. cit.*, p. 130

owing to increased labour costs, and the clay-with-flints soils have fallen from favour compared with the more easily worked and manured soils of the glacial drifts. Some boulder-clay areas it is true, needed chalking owing to surface decalcification, especially where the drift is of Eocene provenance. From our present standpoint, however, the boulder-clay yields a soil distinctly superior to that of the clay-with-flints, and it is probable that the degree of agricultural differentiation between the eastern and western plateaux in Hertfordshire would have been greater

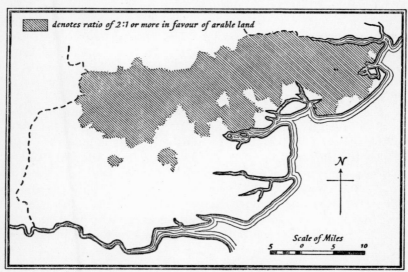

denotes ratio of 2:1 or more in favour of arable land

Fig. 19 Ratio of pasture and arable land in part of Essex

than it is had it not been for the widespread association of the plateau loams with the clay-with-flints.

We may proceed now to inquire how far the soil features outlined above are reflected in the present agriculture of the region. The accompanying maps (Figs. 18 and 19) are based upon the parish returns for the year 1911. Their aim is to show the balance existing in the several parts of the region between arable and pasture acreage. Some difficulty was experienced in selecting the year for treatment as economic causes have substantially modified the arable acreage within the region. There are two possible ways of regarding the operation of the economic factor upon our problem. Agricultural depression might lead to the limitation of arable farming to the regions of optimum conditions, and a map constructed for such a period would then serve to localise these optimum conditions. On the other hand, a period of relative prosperity would presumably lead to the expansion of cultivation to the practicable limits set by soil distribution, and this might conceivably better serve the purpose of defining our regional boundary. Unfortunately the tracing in detail of the changes in arable acreage is a prohibitively expensive research. We have therefore taken the five pre-war years, 1909–13, and using the county returns as a basis selected the year, viz. 1911, which was nearest to the average of the five in respect of arable acreage. We have analysed the 1911 parish returns for an area covering practically the whole of Hertfordshire, Essex south of a line through Dunmow and Braintree, and the contiguous parts of Middlesex.

Reference to Fig. 18 shows that the area of dominant pasture in Hertfordshire (dotted on the map) lies on the London Clay in the south of the county. Its northern limit follows the edge of the drift in the Vale of St Albans with great fidelity. While it is now only just beyond the suburban fringe of London it still retains the characters of a true 'hay country' in large measure. The ratio of pasture to arable acreage is 2 : 1 or more throughout the area so defined, frequently reaching figures of from 4 : 1 to 7 : 1. The Ayot Eocene outlier lying north of the main escarpment also appears as an area of dominant pasture. The area of predominant arable land (shaded on the map) lies chiefly on the boulder-clay plateau east of the Offly-Preston line, which thus contrasts effectively with the chalk-clay-with-flints area of the west of the county. Throughout the area of dominance the ratio is everywhere 2 : 1 or more in favour of arable land, and 3 : 1 may be regarded as the typical figure for the region. It shows, however, two significant departures from the simple accordance with the drift boundary we have traced. In the first place it extends westwards up to and beyond the line of the Lea valley south of Luton. In this we perceive the influence of the high-level plateau loess which, as already noted, is particularly well developed in this area and which extends the area favourable to arable cultivation beyond the line of the glacial drifts. Secondly, there is a decrease of arable acreage towards the south of the boulder-clay tract where several large parishes fall within the 1 : 1 category. In the west the parishes of St Paul's Walden and Knebworth lie chiefly on the clay-with-flints, and no further explanation of their features is required ; but in the parishes of Albury, Benington, and Standon, and the area north of Hertford generally, we see the result of the southerly increasing contamination of the boulder-clay with Eocene débris and the resulting provision of a less suitable soil. The gravels of the Vale of St Albans constitute a transitional area between the tracts of predominant pasture and arable farming, the ratio here being approximately 1 : 1.

In Essex wholly comparable features of distribution are shown. In Fig. 19 the areas are shaded in which arable land predominates (ratio at least 2 : 1). The limit of the main tract runs from the Hertfordshire boundary near Harlow past Ongar and Chelmsford and thence irregularly eastward to the coast in the Tendring Hundred. Thus it clearly follows the margin of the main mass of boulder-clay and of the loams of the Colchester district. The southern dissected fringe of the boulder-clay, particularly in the southwest of the county, falls beyond the area so delimited, and, as in Hertfordshire, the reason is to be sought in the deterioration of the boulder-clay soils within the Eocene outcrop. A marked re-entrant occurs in the edge of the arable country near Chelmsford. This is accounted for by the prevalence of gravelly soils, largely in low-lying situations (parishes of Boreham and Little Baddow, etc.) which are less favourable to arable cultivation than the surrounding boulder-clay. An anomaly less easily explained is the extension of the arable country on to the London Clay northeast of Maldon. There is no special peculiarity of soil other than a slight local lightening by gravel to account for the persistence of arable farming in this area. The London Clay of south Essex was cultivated to some considerable extent before the period of agricultural depression of the late nineteenth century, and here we seem to witness a surviving example of the former condition of these lands. In the same connection we may note a considerable post-war reversion to pasture in the gravel area south of Colchester. It is now predominantly a pasture country, though it appears on the 1911 map as part of an essentially arable area. The contrast noted in passing from this tract on to the boulder-clay near Stanway is as sudden as that seen anywhere in the county.

We may note the areas in which arable land and pasture respectively are strongly dominant. The chief arable areas are three in number. The undissected boulder-clay tract already noted in the west and known locally as the 'Rodings' figures as one of these. It is separated by the London Clay enclave of the Chelmer valley from a long straggling area lying athwart but mainly north of the Chelmsford–Colchester road. Finally there is the clearly individualised area of the Tendring loams. The largest area of strongly predominant pasture lies in the west and includes the environs of Epping Forest. Thence an irregular band stretches eastward along the London Clay outcrop as far as the Blackwater estuary. It is noticeable that at the time in question the London Clay of Essex was not so completely under pasture as that of Hertfordshire. The zone of transition showing approximate equality of pasture and arable acreage, and thus analogous to the Vale of St Albans, is well developed in Essex. As in Hertfordshire it is located partly on gravel outcrops, but large areas of London Clay and of 'Eocene' boulder-clay are also included within it.

Although the foregoing agricultural survey is based upon the figures for a single year, it illustrates the part which physical factors, and more particularly soil factors, play in the life of the region, for the conditions shown on our agricultural maps are still reproduced in all essentials. Moreover, these same factors cannot have failed to exercise an important control on past phases in regional development, for agriculture has been the foremost 'industry' of the area since the days of its first settlement. Accordingly, we may now attempt to assess the importance of the drift boundary as a factor in the historical geography of the region.

An inevitable element of speculation enters into any reconstruction of the earlier phases. In the light of later events, however, it seems wholly reasonable to assume that a beginning was made in the clearance of the drift lands at a very early date. It is virtually certain that the original forest of the drift lands was thinner in character and more easily cleared than that on the heavier clays. On the Ordnance Survey map of Roman Britain an attempt is made to indicate roughly the state of the vegetation of the country at that date, but within our region several points of interpretation seem open to question. In Essex the areas of Bagshot sand figure among the few supposedly free from woodland. Actually the sands are so fine that heathland is practically absent. The soil is not sandy in the ordinary sense and there seems no reason why these areas should not then have supported forest growth as they do today in most cases. The main Chiltern tract is also shown as open country on the Roman map, but in fact the areas of clay-with-flints must certainly have supported oakwood as they do today. It is a surprising feature in several geographical and anthropological treatments of southeast England that the Chalk areas are relegated *in toto*, or very nearly so, to the category of open downland, whereas in actual fact such a character is often conspicuously absent. True downland hardly appears at all on the range of the North Downs east of Guildford save on the very face of the escarpment. Similarly in the Chilterns it is confined to the face for the most part, though it locally extends over the crest. The one locality in which we might expect some extension of treeless country down the dip-slope is in the boulder-clay areas. It must be remembered that in the north of our area the boulder-clay often contains little else but chalk, and its soil, though tending to be somewhat thicker and richer than that of bare chalk, is closely akin to that occurring on the latter. The boulder-clay areas are now so thoroughly divested of even semi-natural woodland that little direct evidence is to be had (Fig. 20). It is not of course contended that the whole drift area was treeless, but merely that certain more elevated tracts near the escarpment edge may have extended the open country to the south and have afforded a natural invitation or entrance to the light woodland country farther down the slope. The boulder-clay plateau of eastern Hertfordshire nearly reaches the line of the Icknield Way between Baldock and Royston. Thence northeastward this ancient trackway parallels the western edge of the boulder-clay all the way to Brandon in Suffolk, and would thus have afforded a natural line of approach to the readily cultivated area, an area barely inferior to the Chalk Marl plain itself over which the trackway runs. It seems, then, not unlikely that Neolithic cultivators made some beginning in the clearing and utilisation of the boulder-clay lands. However this may be, there can be no doubt that the late Celtic or British tribes, and more particularly the Belgae from Gaul, settled in the boulder-clay country.

Several geographical significant facts can be gleaned from our scanty knowledge of the British tribes immediately prior to and after the first Roman invasions. Hertfordshire was shared

Fig. 20   Distribution of woodland in Hertfordshire

between two of these tribes, the Catuvellauni (the Cattyeuchlani of Ptolemy) on the west with their capital at St Albans, and the Trinobantes, whose territory included most of present-day Essex. Eastern or boulder-clay Hertfordshire seems thus to have been grouped with its appropriate natural region, and though we have no knowledge of any precise intertribal boundary, if such existed, it may well have had some relation to the marked change in the character of the country which we have already noted. From Caesar's own account the interesting point emerges that the Trinobantes were the most powerful and prosperous tribe in the district, and among the few that paid a tribute in corn. There is no doubt that the country was effectively farmed.

Further suggestive light is shed by the condition of Britain after the Claudian invasion and settlement. On the map (Fig. 21) the drift boundary is shown in generalised form together with the Roman roads and the location of all evidence of permanent

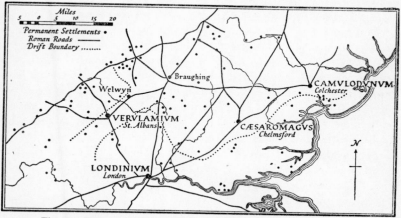

Fig. 21   Roman settlement in the eastern part of the London Basin

settlement.[1] We note particularly the concentration of evidences of settlement within the drift area including all the larger towns except Londinium, and the paucity of such evidences upon the London Clay, particularly in Essex. A further point of considerable interest is the way in which the roads serve to integrate the drift region, suggesting that here lay the most thickly populated portion of the province near to London—a region cut off from London and the Thames valley by thick woodland and marshes. From London three roads lead to the main area of the drift country and only one northwestward towards the distant provincial frontier. All parts of the drift area are effectively linked

[1] The information has been compiled from the Ordnance Survey map of 'Roman Britain'.

by roads, the Colchester road following close to the southern margin, while Stane Street took the axial line from east to west. At or near Braughing in the west-centre of the drift tract six roads met, a greater number than at any of the larger centres except London. Much has been written concerning the factors which controlled the selection of London by the Romans as a base for penetration and organisation. The many good reasons which have been adduced for the selection of a crossing-place here [1] rather than at points east or west can be supplemented by a powerful consideration based upon the relation of the site to the critical drift region. A more easterly crossing, apart from all other difficulties, would have been confronted by a strip of London Clay forest which, though narrow, might well have ranked as a physical barrier. A more westerly crossing might have served for a route to St Albans and the northwest, but in respect of the focal area it would have increased distance and diminished ease of access. It was not, as it seems to us, a question merely of contact with Verulamium and Camulodunum which confirmed the position of London as the growing commercial or administrative centre, but rather of accessibility to a relatively thickly populated agricultural area with well-marked natural boundaries, in which these two towns were important natural foci. London became the capital of its region at a time when its immediate environs were relatively waste and void, and when its function was to act as focus for sub-regions which lay round the periphery rather than near the centre of the main physiographic basin. Prominent, perhaps first, among such sub-regions was the Hertfordshire-Essex driftland. This relation held in Roman times and in large measure remained true for hundreds of years afterwards.

The evidences available for detailed geographical interpretation of the Saxon settlement, here as elsewhere in Britain, are lacking in precision. It is a familiar fact that the invaders came from lands whose geological constitution was precisely similar to that of eastern England in respect of the all-important glacial drifts. A point which has perhaps not been emphasised, however, is that nowhere in northern Europe do the drifts come into contact with so considerable a mass of unrelieved clay as

---

[1] See, e.g. V. Cornish, *The Great Capitals* (Methuen, 1923), pp. 210ff.

is provided by the London Clay. It is fair to surmise that while the Saxon invaders were familiar with every detail of topographic expression and soil character in the drift-covered tracts, the London Clay and the Chalk plateau were unfamiliar terrains. Both its past history and its physical characters marked the drift region as the locale of early Saxon settlement.

A consideration of the present distribution of the settlements within the area affords but a partial and imperfect differentiation between the drift area and contiguous tracts. The unfavourable character and aspect of the London Clay country was relative,

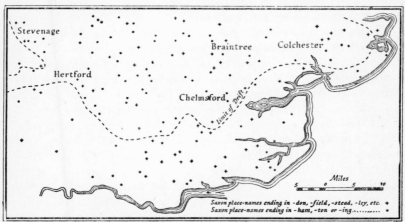

Fig. 22   Distribution of certain Saxon place-names in Essex

not absolute, and large parts of it became settled long before the modern well-sinker had laid the Chalk under contribution to supply its deficiency in underground water. This was particularly so in Essex, where it lay adjacent to the coast. We have accordingly had recourse to the place-names of the area as a means of throwing light on the early progress of the Saxon settlement. It is widely accepted by students of the subject that settlements whose name has the termination *-ham*, *-ton*, or *-ing* are earlier than those of a descriptive or topographical character, with such endings as *-don*, *-field*, *-stead*, *-ley*. Fig. 22 shows the distribution of both sets of terminations. The first is seen to be concentrated within the drift area, and notably in the west, where the boulder-clay shows its widest expanse. This is one

195

of the few cases in which such settlements were not in river valleys. Those beyond the drift-limit lie close to it, or else are situated within the area of the Thames gravels. The London Clay area is almost without evidence of settlement at this early date. The same map (Fig. 22) also shows the distribution of the second set of terminations. As would be expected, there is evidence of further settlement in the northern areas, but there is notable evidence of this presumably later phase of settlement on the London Clay. At the risk of begging the question it may be urged that the high geographical probability of the suggested order of settlement affords some confirmation of the value of place-names in such studies as this.

From the history of Saxon days several other facts emerge of which a geographical interpretation may be given. The diocese of London constituted at the beginning of the seventh century in the charge of Mellitus excludes from its limits a large part of the true Chiltern country. From the 'Tribal Hidage', a document assigned to the same period, we learn of the existence of a tribe—the Chilternsaeta—who inhabited this country which there is every reason to think was still thickly wooded. These people are believed to have been of British origin, a supposition supported by the high degree of nigrescence in the present population.[1] The main interest of the matter in the present connection is the strong suggestion conveyed that Saxon settlement was concentrated upon if not almost confined to the eastern drift plateau, while some element of the former population lingered in the wooded Chiltern country, a relatively negative area. Mackinder suggests that the isolated region of nigrescence owes its existence to its position between the extreme limits of Anglian and Saxon (West Saxon) penetration. This is no doubt true, but it seems permissible to urge as a further factor the change in the character of the country which takes place at the Offley-Preston line. Although the West Saxon and Anglian immigrations had spent their force in this area it remained open to penetration by the East Saxons except in so far as the route from the east encountered physical barriers. Although the Offley-Preston ridge is not in the narrow sense a barrier, as a limit between regions

[1] H. J. Mackinder, *Britain and the British Seas* (O.U.P., 2nd ed. 1907), p. 182

of easy and relatively difficult utilisation of the soil, its importance is not to be gainsaid.

The gradual growth, conflict, and adjustment of the several Saxon and Anglian kingdoms was guided in no small degree by geographical factors, and in this phase again we perceive the influence of the Offley-Preston line. The western portion of Hertfordshire formed part of Mercia at one stage, while the eastern drift area again followed its natural geographical affinities, falling within the confines of Essex. Little is known in detail about Saxon Essex ; it seems likely that it was not fully settled until after the establishment of the Angles in Suffolk. It has been contended by some that the existence of a strong fortified centre at London blocked the route up the Thames valley, and directed penetration up the smaller estuaries to the north which, as we have seen, lead up into the heart of the drift country. All the actual Saxon remains have been found on or near the boulder-clay tract, and chiefly in its eastern or seaward portion. Beddoe [1] was of the opinion, from the study of the present population, that a considerable British element survived in parts of Essex, chiefly, we may surmise, on the clay lands of the south. The critical area in respect of the Saxon penetration is unquestionably the watershed between the Chelmer and Roding valleys. Eastwards the valleys lead to the coast, but the country to the west is not so easily penetrated from that quarter. Hence it is likely that the immigration did not extend into the area of the Rodings, the Stort valley, and eastern Hertfordshire until a relatively late date. Nevertheless, as we have seen, the western part of the boulder-clay plateau was intrinsically a more favourable area than the east, and ultimately it became more thickly settled.

The critical area of the drift margin in Hertfordshire again came into prominence during the Danish invasions, standing out as a zone of oscillating frontier or a march land. Danish influence at its maximum extent penetrated into western Hertfordshire, but the Treaty of Wedmore (878) between Alfred and Guthrum fixed the boundary of the Danelaw at the line of the river Lea from its source to its mouth. Thus the whole of the drift country was included within the Danish realm. It is of course evident that the Offley-Preston ridge, though a salient

---

[1] J. Beddoe, *The Races of Britain* (Arrowsmith, 1885), p. 254

physical feature, is hardly sufficiently definite to form the basis of an actual frontier, quite apart from the fact that as a defensive line it would favour the western or Chiltern power unduly. There is no continuous or considerable stream through the Hitchin Gap, the headwaters of the Beane and the Hiz overlapping. Thus the frontier came to be fixed at the first readily demarcated line west of the true regional boundary. During the later phase of re-conquest from the Danes, when the shire system was extended, the boundary of Hertfordshire was shifted eastward across the drift country to the next marked physical line—that of the Stort-Cam depression. Essex, though it fell within the Danish realm for a long period, was not thickly settled by the Danes except in the coastal and riverine tracts and in the north on the boulder-clay. Their sojourn in the latter area set its mark upon the country and served further to emphasise the contrast with surrounding lands by the introduction of characteristic differences of land tenure, notably the extension of a class of freeman holders.

It is a striking demonstration of the co-operation of physical and historical factors that three successive waves of immigrant peoples, those of the Belgae, the Saxons, and the Danes, should have settled in the same tract and thus have served to con-solidate the distinctiveness of its regional characters. It is true that each immigrant wave was guided along the same natural routeways into the country and thus tended to reach the same regional goals, but their ultimate settlement was undoubtedly conditioned by the inherent richness of the drift-lands.

A wealth of further evidence bearing upon the value and desirability of the drift-lands can be gained from the records of the Domesday Survey.[1] Conversely it is perhaps not too much to suggest that the interpretation of these records would gain much from an explicitly geographical basis and treatment, such as does not seem hitherto to have been accorded them.

A striking fact which emerges from the Domesday Survey of Hertfordshire is that the vills on the eastern side of the county were divided among large numbers of free tenants or small

---

[1] For information upon this subject we have depended upon the Victoria County Histories for Hertfordshire and Essex (chapters on Domesday Survey by J. H. Round).

holders (socmanni) in the reign of Edward the Confessor. After the Norman invasion they became a depressed class, and much reduced in numbers, their lands being merged in the Norman manors. Their distribution in 1066 shows highly significant features, for the number falls off distinctly from east to west, and the westerly fringe of the sokelands stretched from the neighbourhood of Lilley (west of Offley) to Hoddesdon, and two-thirds of the total number of sokemen (excluding those of the Manor of Hitchin) were to be found east of a line drawn from Royston to Hertford. The plain relation between this distribution and that of the glacial deposits as described above has not, to the best of our knowledge, been pointed out previously. Maitland's [1] opinion, based upon his study of the Domesday records, that this area was 'richer and more populous' than the regions to the west is entirely in accord with the arguments presented in this paper, but it is probable that we see in the distribution of the sokemen no more than a legacy from the Danish occupation of the district, an occupation which, as we have already seen, was effectively limited to the drift-lands.

That Hertfordshire had for long been an arable county is evidenced by the fact that at the time of the Domesday Survey its wealth was derived mainly from the plough, though it was supplemented in order of importance by water-meadow, pasture, woodland, fisheries, and water-mills. That a transition from a pastoral to an arable country was observable in passing from south to north is also strongly suggested. Thus we read that the manors of Hoddesdon, Cheshunt, Wormley, Broxbourne, and Amwell had sufficient hay to feed all their plough teams, the two last, indeed, having an excess for sale. Thundridge and Standon, farther north, had just sufficient for feeding purposes, while Braughing had hay for only three of its teams. Other manors mentioned as having a deficiency of pasture are Westmill, Datchworth, and Digswell.

From the Essex record further interesting facts emerge. Thus, it seems a legitimate inference that little pasture was yet available within the confines of the London Clay forests, for the very considerable number of sheep in the county, estimated roughly at the time of the survey as 18,000, were grazed chiefly

---

[1] F. W. Maitland, *Domesday Book and Beyond* (C.U.P., 1897), pp. 22–23

on the coastal marsh strips, at Southminster, Burnham, Dengie, etc., and the associated cheese-making was also centred on the marsh lands. Many inland parishes had pasturage rights on the marshes, and tracts such as Foulness Island and Canvey Island were broken into a patchwork of pasturage divisions attached to inland parishes.

Some idea of the distribution of forest can also be gained. The southwest of the country was thickly wooded and a wide belt of woodland stretched southward from Ongar to Brentwood. The manors particularly mentioned as being rich in woodland are Blackmore, Mountnessing, Hutton, Little Warley, Bulphan, Orsett, and Cranham. A belt of woodland stretched southward from near Hatfield Peverel to Woodham Ferrers and hence southeastwards to Purleigh and Latchingdom. In the north of the county in the Hundred of Hinckford woodland is recorded as scarce. Thus the fact is clearly brought out that the forest was located chiefly on the London Clay area. It is not, however, to be supposed that the boulder-clay area was devoid of forest, for parts of it are specifically mentioned as having a large extent of woodland. There is, however, a suggestion that its amount decreased from the south to the north, and one is tempted to correlate this with the increasingly chalky character of the boulder-clay in this direction. Considerable areas of woodland were cleared in Essex between 1066 and 1086, no doubt in large part for purposes of cultivation. There are definite references in the early charters of Essex and other areas to 'assarts' or clearings for cultivation. It is significant that in Essex the greatest clearing took place at Takely, Ugley, Thaxted, Wimbish, Notley, etc., all situated on the boulder-clay terrain.

Turning lastly to the distribution of the manors noted in the Domesday Survey (Fig. 23), we at once perceive a notable concentration in northwest Essex and eastern Hertfordshire. In Hertfordshire the maximum concentration occurs within a rectangle whose four corners are marked by Hitchin, Hertford, Bishops Stortford, and Barkway. Within this area they occur on the level boulder-clay surface, often occupying the sites of what are today plateau settlements. South of the line Hertford— Much Hadham—Bishops Stortford, where the boulder-clay sheet is broken by the valleys of the Rib, the Ash, and the Stort, there is a marked falling off in the number of manors, which are here situated in the valleys well above the flood plain. Within the area of the Chiltern plateau manors are relatively few. They occurred almost exclusively in the valleys, the broad clay-capped interfluves being almost devoid of evidence of settlement. The Vale of St Albans has two parallel lines of manors along its northern and southern margins respectively. Except for one at London Colney there were no manors in the centre of the Vale.

Fig. 23   Distribution of manors in Essex noted in Domesday Book

In Essex the area of the Thames gravels is thickly covered with manors, but the London Clay belt in the west shows but a sparse distribution. Farther east, in the area north of the Crouch and between the Blackwater and Colne estuaries, the London Clay area, here adjacent to the coast and penetrated not only by the major estuaries but by innumerable smaller creeks, was thickly settled. There is a marked difference between the eastern and western portions of the boulder-clay tract in respect of the distribution of manors. In the east they follow the valley ways, avoiding on the whole the interfluve areas. In the west beyond the Chelmer valley we encounter the eastern side of the densely settled district already noted in Hertfordshire. The broad belt of manors which extends thence southward down the Roding valley indicates some northward penetration of the area along this route. The map also suggests the importance of the

Lea and its many-branching headstream system of waterways as a means of access to the thickly settled district. We may recall that the Roman road to Braughing crossed the difficult high and dissected country west of the Lea valley, but in later days the main waterway artery had attracted settlement down to the broad gravel and loam-covered valley floor and doubtless led to the founding of the modern roadway which follows the valley line. Braughing was the 'Hertford-Ware' of the Roman system in a day before water-transport had come into its own.

Concerning medieval and modern times little need be said here. While the superior value of the drift-lands continued to operate as a factor, the social and economic history of the region shows the working of so many other factors that it is no easy matter to separate the geographical and the political elements of the problem. And the subject certainly demands more detailed treatment than we can accord to it here. In particular the growth of London as a true regional capital has transformed the regional setting almost beyond recognition, dominating and obliterating local markets and enhancing the value of the London Clay lands surrounding the capital. It is, however, interesting to note that in the thirteenth and fourteenth centuries when the average price of land in Hertfordshire was about fourpence per acre, higher prices, up to sixpence and more, ruled in the boulder-clay area, the zone of high prices extending from Langley to Sawbridgeworth.

It is also significant that of the smaller settlements which rose to local dominance as market towns, a large proportion lie on or near the drift boundary, and are thus situated in the zone of transition between essentially contrasted regions. We may instance Hitchin, Hertford, and Ware, Ongar, Chelmsford, Witham, Maldon, and Colchester as examples of this class. In most cases other factors have co-operated in securing the relative advantage of the site ; but the relation to the drift boundary is one of the most important features and one of the few that the sites have in common. The high degree of natural nodality of Hertford and Ware contributed to their past importance, an importance maintained in spite of mutual strife and the rivalry of London. The region so effectively integrated by the head-waters of the Lea converging on these towns had during past centuries and still retains in some measure the status of a separate sub-region, one of high farming and closely associated with the malting industry. The urban group at the western end of the Vale of St Albans, comprising chiefly Watford and Rickmansworth, is closely analogous in position so far as the morphology of its site is concerned ; but it shows considerable contrasts in history and present development. These are largely due to the relative poverty of the surrounding country in respect of soil. As a consequence of this and of the easier access to London, Watford and Rickmansworth do not possess the character and atmosphere of local centres so strongly as Hertford and Ware, and the shadow of the Metropolis has fallen more heavily upon them.

12

Some Aspects of the Saxon Settlement
in Southeast England

Considered in Relation to the Geographical Background

IT CAN hardly be gainsaid that the Saxon settlement was the making of England in the geographical as well as in the historical sense. Notwithstanding considerable contributions to our knowledge of the Roman and prehistoric phases and the radical modifications brought about in later medieval and modern times, it remains true that the cultural landscape of much of the English Plain shows mainly the impress of this major settlement of the region. No understanding of English 'rural' geography is possible without reference to the circumstances under which its ground-plan was established. Too often the attempt is made to explain village sites in terms of their present geographical setting. To some extent, admittedly, geographical controls are operative, but, in general, very false ideas of geographical causation must derive from ignorance or neglect of the conditions which held in the phase of settlement itself. The subject has another claim upon the attention of geographers. It is one to the discussion of which geography has very definite contributions to make, though these have sometimes been ignored or under-rated by historians and archaeologists.

The older and more familiar interpretation of the Saxon Conquest and settlement, followed by J. R. Green [1] and others, envisaged the landing at various points on the coast of a number of separate marauding bands and the establishment thus of the nuclei of kingdoms, which ultimately with expansion of area came into contact and conflict in the phase of the so-called 'struggle for supremacy'. This simple and essentially reasonable conception has latterly given ground to another historical theory

[1] J. R. Green, *The Making of England* (Macmillan, 1882)

which represents the initial act of conquest as a concerted operation on a wide scale, leading to an early overrunning and sub-
jugation of a large part of Romanised Britain—then in a phase
of chaos and dissolution. The triumphant march of the invaders
is believed to have been checked at the battle of Mons Badonicus,
after which, in the moment of defeat, the wave of invaders
became disrupted and 'crystallised' into kingdoms. The evidence
for the former theory was largely derived from the Anglo-Saxon
Chronicle. This formed the basis, for instance, of the coherent
and essentially credible narrative presented by J. R. Green in
The Making of England. The latter phase of opinion was based
in part upon the evidently garbled and doubtfully authentic
character of the earlier entries of the Chronicle. The more
nearly contemporary statements of Gildas and certain Continental
workers have been deemed a preferable and more dependable
source of information. Chadwick [1] supported, if he did not
initiate, the newer theory, using the Danish invasions as an
analogy for a planned and concerted attack. Bede's reference
to the 'imperium' or 'bretwaldaship' held in turn by the Anglo-
Saxon kings has also been pressed into the service of the argument,
as a supposed survival of the earlier invasion unity.

Into the historical merits of this second theory it would not
be fitting to enter here. The reader will find it treated somewhat
fully by Oman,[2] in so far as the earlier documentary sources are
concerned, and this writer further remarks that the theory of an
early widespread invasion receives some support from the well-
known fact of the early destruction or decay of the Romano-
British cities, as attested by archaeology. The theory has recently
been criticised by Lennard,[3] who adduces reasons for preferring
the older theory. He at least emboldens us to remark that the
theory of an early 'concerted rush' is somewhat lacking in
geographical probability. Much of the country crossed was still
forested and difficult, if not virtually impossible, to settle quickly.
Nevertheless, it cannot be pretended that geographical arguments,
as such, can bear much weight in the matter of this supposed
early overrunning, much less disprove the idea. In view of the

[1] H. M. Chadwick, *The Origin of the English Nation* (C.U.P., 1907)
[2] C. Oman, *England Before the Norman Conquest* (Methuen, 1929)
[3] R. V. Lennard, *History* **18** (1933–4), 204

survival of a network of Roman roads giving ready access to the west, there is nothing inherently improbable in the idea of penetration of relatively large war-bands to within sight of the 'Western Sea'.

What is far more relevant, we submit, to the geographer's consideration is the distinction which can and should be drawn between 'conquest' and settlement. The geographer may accept the idea of an early conquest on the evidence submitted, but he must, we think, repudiate strongly the improbable theory that the kingdoms 'crystallised' in any sort of ready-made form during the period of disruption, or indeed in any other phase. It seems not unreasonable to take exception to such a notion of the origin of kingdoms, as also to that other and related conception which speaks of them as 'carved out'. Under conditions such as those of the Saxon settlement there can be little doubt that kingdoms *grew* by a process of arduous piecemeal penetration of groups of settlers who owned an allegiance or came to own an allegiance to a reigning house. If 'conquest' there was, settlement had none the less to begin and to take its normal gradual course in later stages. It is with settlement as affording a definite contribution to the cultural landscape, not with conquest as such, that the geographer is concerned.

It is difficult to bring any geographical viewpoint to bear on the vague and inconsequent statements of Gildas and the other earlier writers. Though the events of which they speak can be approximately located in time, their spatial relations are indefinite to a degree. Geography gets its first chance of a voice in the problem when we have recourse to the study of dateable Saxon antiquities and of place-names. Here are determinate, though necessarily incomplete, distributions which, in effect, define areas of settlement and suggest routes of penetration, and here a comparison with physiographic conditions becomes fruitful. But the geographer would wish to go further than this and to construct to the best of his ability, quite independently of the archaeological or historical evidence, the contemporary physiographic and vegetational background. In the light of the long discussions of past decades, he will probably not claim that the features he portrays possesss any 'compelling' quality. They represent the field of physical opportunity as it presented itself for the cultural modification of the settling people.

It cannot be too strongly urged that one of the main contributions the geographer can make to the study of any period is a synthesis of the physique, designed to bring out distinct minor regions with clearly traceable boundaries. In France and certain other parts of Europe, the existence of such regions has forced itself upon the attention of the peasant, and a complex minor regional nomenclature exists in the vernacular and can be adopted with slight modification by the geographer. In Britain equally well-marked regional subdivisions exist and traces of a '*pays* nomenclature' can be found, but it is much less extensively developed than in France. In the construction of a map of regional subdivisions for southeast England, we are primarily concerned with (1) soil, (2) natural vegetation, (3) water supply, and (4) texture or pattern of relief and drainage (*not* altitude, as some treatments would seem to imply). Neither a geological nor a contoured map is fully sufficient for our purpose. The task lies essentially in defining the lines, or narrow zones along which there is a clearly recognisable change in the physical landscape, and probably also in the modern cultural landscape ; which reflects some at least of the contrasts effective at an earlier date.

In Fig. 24 we present a tentative regional sub-division of southeast England on the bases indicated. In spite of difficulties of construction, all boundaries shown at least satisfy the test suggested above—i.e. they are visible, and therefore 'mappable' regional boundaries. In so far as soil, morphology, and water supply are concerned, the several regions possessed largely the same features in Saxon times as they do today. Into the reconstruction of the original vegetation conditions some element of hypothesis necessarily enters. We would point out, however, that there is no reason to suppose that rainfall, the most likely variable, was appreciably different in early Saxon times, though the sixth and seventh centuries may have been drier and the Romano-British phase wetter than the present era. If this be true, the margin of error in reconstructing past vegetation conditions is considerably reduced.

The heavy clay areas, formerly thickly forested, are first delimited. They include parts of the outcrop of the London Clay, the Weald Clay, and the Wadhurst Clay of the Rother valley. The western Wealden area differs from the 'vales' of Kent, Surrey, and Sussex, showing a markedly more accidented topography. As a separate category, certain 'drift-lightened' clay lands are distinguished. These include dissected plateaux, whose gravel-capped 'tops' afforded sites for characteristic

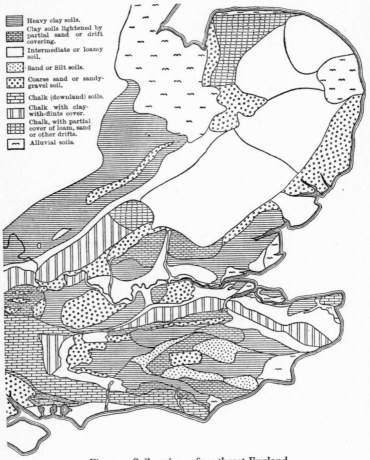

Fig. 24 Soil regions of southeast England

plateau settlement, as well as lowland areas like that around Hadlow, south of Maidstone, where river-drift diversifies and modifies the clay flats. Here also are included certain tracts, such as the Vale of Holmesdale, and parts of the Kennet basin and of the central Weald, where small sand and clay terrains occur intimately intermingled, and where accordingly the process of wash on the slopes gives rise to numerous ill-defined loamy streaks.

Next in logical order come the areas of loamy or 'intermediate' soils. They comprise several variants, and are perhaps best generally characterised as the equivalents of 'les sols de cultures faciles' of Vidal de la Blache. We have sought to define the areas in which 'brickearth,' lithologically comparable with true loess occurs extensively, whether at high or at low levels. Such areas include the valley loam-terrains of the Thames and its tributaries, the Norwich and Southend loam regions, and the 'Sussex Levels'. It is essential to emphasise, however, that closely similar soil conditions are dominant in the following regions widely separated both geographically and geologically : (1) The boulder-clay plateaux of East Anglia and the London Basin, especially in Norfolk, Essex, and east Hertfordshire. The 'Chalky-Kimmeridgian' boulder-clay of Suffolk gives rise to somewhat heavier soils ; the area of the heavier soils is indicated on the map, but not classified. (2) The Lower Tertiary belt in the southeast part of the London Basin. It is the Thanet beds that are of chief importance here, and it should be noted that their relics modify the soil over a belt of variable width at the foot of the Chalk dip-slope, here included as part of the same soil region. The main portion of the characteristic Lower Tertiary belt ends near the southeast outskirts of London, but it continues as a narrower strip farther west into Surrey. (3) The scarp-foot or 'Icknield zone' along the western foot of the Chiltern Hills and the East Anglian Heights. This is sited on the marly Lower Chalk. Similar conditions exist in the western Weald where the broad Upper Greensand bench contributes its quota of readily arable soils. Comparable conditions exist indeed at the foot of the Chalk escarpment throughout the Weald, but, except in the west, the belt in question is not wide enough to be shown on the map. (4) The area of the East Kent Hythe beds (part of the Lower Greensand) and that situated on the Bargate beds and Loamy Folkstone beds of west Surrey.

The essential unity in geographical characteristics of these several regions is apt to be masked [1] by the accidents of a geological

[1] S. W. Wooldridge and D. L. Linton, *Antiquity* **7** (1933), 297

nomenclature designed to serve stratigraphical not geographical ends.

In the regions of sandy type, we have made a twofold division, taking first the regions of fine sands or silts in which the soil does not invariably favour heathland conditions, but is retentive of moisture, and suitable either for arable or grass land. Contrasted with such areas are the true sandy (or gravelly) heathlands, prominent among which are the following : The New Forest area, the 'Bagshot plateau', the Lower Greensand areas of the western Weald, the higher parts of the central Weald, the 'Blackheath plateau' of southeast London, the Beaconsfield gravel plateau, and the contiguous tract of glacial sand and gravel which occurs in the Vale of St Albans. Farther to the north and east we have the similar gravelly tract in Essex, between Chelmsford and Colchester, the Suffolk 'sandlings', the Breckland and the Norfolk and Bedfordshire Greensand regions. It cannot safely be asserted that all of these tracts were in the condition of heathland in early times, but the woodland cover must have been light ; biotic factors, including interference by man, are partly answerable for their present condition. Moreover, soil and water-supply conditions alone mark out these regions as fundamentally distinct. It may be remarked that the topographic and cultural expression of such coarse sand and gravel terrains depends very largely upon their form and altitude, which control their relation to the water-table. Most of the regions listed above are of plateau character, with their surface well above the water-table ; in this respect they contrast with the valley—or terrace—gravel regions, which rarely show dry and acid soil conditions.

Finally, we have distinguished the areas of true Chalk downland, separating them from those other important areas in which a cover of clay-with-flints, or lighter drift, masks much of the surface. A partial cover of clay-with-flints exists on the culminating points of all the Chalk areas, but in Sussex and Hampshire and over most of the Wiltshire and Berkshire Downs, it is of insignificant extent. In the plateaux of the Chilterns and the North Downs very different conditions prevail ; there is an extensive development of clay-capped 'top' which must have been forested in early times and is indeed still thickly wooded. The eastern end of the Chiltern plateau is shown as extensively loam-covered, and the Chalk area of west Norfolk is grouped as of similar type. It is probable that the eastern end of the North Downs should be placed in a similar category. In these areas, plateau brickearth or some kindred deposit, rather than clay-with-flints, covers much of the surface, and from the standpoint of soil conditions these regions have affinity with those of the loamy or 'intermediate' class mentioned above.

Areas of present alluvial marshland, whether fluviatile or marine, are indicated in conventional manner in Fig. 24. These were formerly largely water-covered, especially following the slight submergence which seems to have taken place during the Roman occupation.

The limitations of the method pursued in constructing Fig. 24 are not far to seek ; they are inherent in any attempt at generalisation. In so far as the regions are soil regions, it is soil 'tendency' or mode which is the basis of division. In respect of the degree of detail recognised in soil variation, the map is intermediate between those embodying the results of actual soil-survey (yet to be made) and those which group the whole of the soils of the English Plain under the unhelpful title of 'Brown Forest Soils slightly podsolised'. Some of the regions recognised are more homogeneous than others, and it has proved neither possible nor desirable to apply a rigid and invariable criterion of delimitation. The result is a tentative synthesis subject to future improvements and requirements. Nevertheless, it is hoped that it presents, on the whole, a better picture of the physiographic background than could be obtained in a map of this scale with either simple geological or simple orographic data.

In investigating human reactions to this mosaic of minor regions, we might, for instance, attempt to *deduce* the probable mode and extent of utilisation by man of the contrasted regions. Such a course does not recommend itself, for the method of geography is synthetic description, not deduction. It will be preferable, as always, to strive to answer the definite question, how and when the several regions were used. Again, it would be artless to expect any exact correspondence between 'human' and physiographic regions. The human region in any given historical phase tends to be compounded of parts of contiguous physiographic regions, with its centre near their common meeting-place. Nevertheless, there is a certain sense in which the regions distinguished above can be grouped as positive or negative, favourable or unfavourable, from the standpoint of an immigrant agricultural race with a limited control over its environment.

The briefest survey of pre-Saxon conditions in southeast England is sufficient to assure us of the reality of physiographic controls of human distributions from the very earliest days of decipherable settlement. It is true that there have been noteworthy changes in 'geographical value' more than once. Mesolithic man alone, till the days of the suburban builder and the modern military camp, found the sandy heathlands favourable to his activities. The outstanding tendency of the earlier periods was the so-called 'valley-ward' movement—the change from the 'prehistoric' to 'historic' mode of settlement. This can be more accurately characterised as a movement towards 'better' soil areas—'better' at least in relation to the gradually increasing power of agricultural technique. This great change in the siting of settlement and agricultural land began not later than the Bronze Age, and the Saxon settlement marks its virtual culmination.

From the geographical standpoint there are three distinct phases in an episode such as the Saxon settlement. The study of the initial phase involves a consideration of those geographical features which controlled access and entry by the immigrant people and their 'skeletal' infiltration by way of natural lines of communication. Secondly, there follows rapid expansion and colonisation from a few early settled tracts. Finally, the settlement plan fills out and assumes for the time being a relatively static condition. These three phases may for brevity be called the Entrance, Expansion, and Terminal phases. Apart from the scanty and ambiguous documentary evidence, we have the following sources of information as to the progress of the Saxon settlement and the establishment of village sites : (1) the distribution of Saxon burial-grounds of the pagan period as studied by Leeds and others ; (2) the distribution of place-names, especially those of the group deemed early, with the terminations -*ing* and -*ingham* ; (3) the Domesday Survey, with its records of vills. It is evident from recent researches that the study of field-systems and of tenurial custom generally may also throw some light on the settlement, but the information does not exist at present in a 'mappable' form, and so does not lend itself readily to the method of treatment here employed.

In a broad sense, it appears to us legitimate to take the mapped evidence from the three above-named sources as representative of the Entrance, Expansion, and Terminal phases of the settlement, respectively. Such a course is not free from objection : the first two categories of information overlap chronologically to some extent, and neither really takes us beyond the early

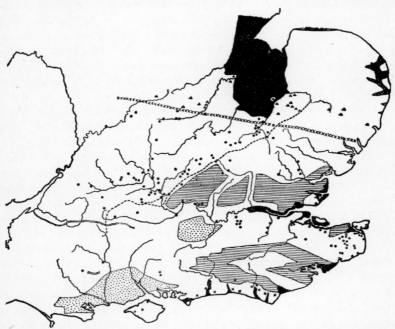

Fig. 25 Map to show distribution of early Saxon burial grounds. (After Leeds—modified.) West and East Saxon sites shown by dots, South Saxon by crosses, Jutish by circles, and Anglian by triangles. The Saxon-Anglian cultural divide is indicated (line of crosses), also the Icknield Way (dotted line) and the Wansdyke (double line). Marsh and Fen in solid black, heavily forested clay lands in horizontal lines, heathlands dotted

stages of the settlement. It will be indicated, however, that the place-name method is capable of some extension.

Fig. 25 shows the distribution of early Saxon burial-grounds, together with certain of those showing Anglian cultural affinities and those marked by the very definite differentiae of the Jutish culture. We have not thought it practicable to show the whole of the physiographic background as portrayed in Fig. 24, preferring to single out certain of the more significant regional features, which facilitate a comparison with the more detailed map.

Distributional evidence gives no reason to doubt the Jutish landing at Ebbsfleet. We note that the burial-grounds are widely distributed on the relatively open Chalk plateau of east Kent, where the dominance of loamy soils has been noted. The westward extensions are along the important Lower Tertiary zone of the marsh-edge, the line followed by Watling Street, and also to a limited extent along the Lower Greensand belt. It is noteworthy that these early evidences are confined to the part of Kent lying east of the Medway. The separate settlement in the Hampshire region is also evident.

The South Saxon settlement appears as based on the loam region of the Sussex Levels, with extensions through the river gaps to the scarp-foot zone beneath the Downs. In Essex, the evidence suggests penetration by the Blackwater estuary, which affords a short crossing of the belt of London Clay forests and ready access to the drift-lands. There is also some sign of occupation on the Southend Loam plateau.

The most striking feature of the map, however, is the concentration of burial grounds along the Icknield zone and the country to the north of it, coupled with the almost complete absence of contemporary evidence in the region of the Hampshire Chalk. This is the basis of the claim made by Leeds [1] that the traditional account of the settlement of Wessex is seriously at fault. In view of this distribution and the fact that the absence of evidence in Hampshire and Wiltshire is most unlikely to be due to collection failure, Leeds first favoured the view that the main West Saxon entry was by way of the Thames, a suggestion previously made by Sir Henry Howorth. The earlier burial-grounds in *west* Kent and Surrey are certainly of West Saxon affinity, and mark some of the earliest recorded settlements of the English peoples. Latterly, Leeds [2] has preferred to envisage the early West Saxon penetration as taking place along the line of the Icknield Way, from the rivers of the Wash, before the

---

[1] E. T. Leeds, *The Archaeology of the Anglo-Saxon Settlements* (O.U.P., 1913)
[2] E. T. Leeds, *History* 10 (1925), 97 and *Brit. Journ.* 13 (1933), 229 ; but see also M. W. Hughes, *Antiquity* 5 (1931), 291

spread of the Middle Angles across the path. Certain geographical comments seem relevant here. The traditional account of the West Saxon entry from the south is not only somewhat insecurely based historically and at issue with the archaeological evidence, but presents curious geographical problems. The story of the Anglo-Saxon Chronicle, even as elaborated and rationalised by J. R. Green, represents the West Saxon expansion as halting for many years in what was perhaps the most open and readily penetrable country in southeast England. While it is not for the geographer to say that such a state of affairs is impossible, it does not commend itself as at all likely, in view of the progress made in more difficult country elsewhere. Moreover, it is to be noted that the Southampton Water entry leads through the broadest part of the forested Hampshire basin; the spread of earlier peoples to Salisbury Plain, from points farther west on the coast, was opposed by much less serious obstacles. None of these possibilities, however, preclude the possibility mentioned by Chambers,[1] that a small band of raiders under 'Cerdic' landed near Southampton Water and finally settled down with the main body of West Saxons in the upper Thames region.

The suggested Thames entry has much to recommend it, and there is no reason to suppose that either London itself, or riverside forests above London,[2] opposed any insuperable barrier to penetration. In regard to the former, there can be little doubt that A. Major's neglected suggestion that London could be by-passed at high water is reasonable enough.[3] There was undoubtedly some submergence of the Thames-side lands in the second century, and consideration of the geological conditions supports the view that direct passage from Limehouse Reach to Lambeth Reach, south of the main Thames channel, would have been possible. If so, London lost for the time being much of its significance as a defensible riverside site. The Thames flood-plain above London is margined almost continuously with high and dry 'gravel-ways' which must have assisted up-river penetration. Nevertheless, the relative scarcity of burial-grounds along the middle Thames and the apparent failure to utilise the

[1] R. W. Chambers, *England Before the Norman Conquest* (Methuen, 1923), p. 92. See also O. G. S. Crawford, *Antiquity* 5 (1931), 441
[2] E. T. Leeds, *The Archaeology of the Anglo-Saxon Settlements* (O.U.P., 1913)
[3] A. Major, *Proc. Croydon Nat. Hist. and Sci. Soc.* (1920), 1

valleys of the northern tributaries suggest that the Thames entry was comparatively unimportant at this stage.

Though the Thames itself was no doubt a practicable waterway, we are disposed to believe that Leeds,[1] Baldwin-Brown,[2] and others have over-estimated the importance of ingress by water during the earlier Saxon phases. The riverside strings of burial-grounds are admittedly a striking feature, but it is to be remembered that the flood-plain margins, where underlain by gravel, offer other attractions, such as dry foundations and available water-supply, quite apart from possibilities of access by water. The force of this consideration is seen when we recall that early burial grounds occur close to the Warwick Avon and its tributaries, yet it has never been suggested, as far as we know, that they are attributable to invaders moving up from the Bristol Channel. Some element of land movement is here clearly implied. Such land movement is implicit in Leeds' theory of penetration by the Icknield Way, and we may perhaps add a comment of the unique qualities of the Chalk scarp-foot zones in general and of the Icknield zone in particular.

We have already noted the favourable soil conditions which exist in a belt of variable width at the foot of the several Chalk escarpments, whether the Upper Greensand is present or not. In the case of the Chiltern escarpment, the Chalk Marl outcrop is particularly wide, owing partly to glacial erosion of the scarp-face, as in Cambridgeshire, and Bedfordshire, and partly to other and unknown causes, as in Buckinghamshire, Oxfordshire, Berkshire, and Wiltshire, where the influence of ice cannot so readily be invoked. But in addition we have to take account of the scarp-foot spring-lines (sometimes double in the case of the Chiltern foot), perhaps the most consistent and continuous water-yielding lines in the country. Again, the scarp-faces formed well-marked 'guiding features' into the heart of a forested country, a function assisted by the fact that whatever the condition of the scarp-crest, the face was probably clear of forest from the earliest times and was probably the only tract so characterised. Much emphasis has been laid upon the 'crestway' tracks of the more open downland country in the south, and by analogy the idea

---

[1] E. T. Leeds, *The Archaeology of Anglo-Saxon Settlements* (O.U.P., 1913)
[2] G. Baldwin-Brown, *The Arts in Early England*, vol. IV (Murray, 1915)

has grown up that the crests of the North Downs and Chiltern Hills were similarly used. No-one with knowledge of the ground could support such an idea for a moment. It is the scarp-foot zones which were significant. They were natural avenues or glades, inviting the construction of tracks of easy contour and dry foundation which could be followed even in bad weather or semi-darkness with the help of the numerous aligned projecting salients of the scarp. They were well watered and threaded some of the best areas for cultivation in the whole countryside. Of all the scarp-foot zones, the Icknield zone is the most important. It retains its essential character of a more or less dissected Chalk Marl platform from Hunstanton to Devizes, a distance of more than 150 miles. Beyond Devizes it is replaced by the Upper Greensand bench. The analogous Pilgrim's Way zone of the northern Weald has fulfilled a kindred historic rôle, but has been subordinate in importance to the 'Watling Street zone' at the foot of the dip-slope. In the South Downs, for evident reasons, longitudinal movement has followed the scarp-crest ; the scarp-foot road is disjointed, and appears to be a composite trackway made up of separate portions linking later Saxon settlements.

We may now pass to a consideration of place-names in their bearing on early settlement. Fig. 26 presents us with a distribution which we may regard as affording a picture of the earlier part of the 'expansion phase'. On it are marked the sites of place-names ending in *-ing* and *-ingham*, listed by Ekwall [1] in his well-known study. These place-names are held to imply an old 'tribal, clan, or folk system' which died out during the later phases of the settlement, and their claim to be regarded as early is admitted by all place-name students. The first to demonstrate the importance of the group of place-names ending in *-ing* was Kemble,[2] who as long ago as 1849 listed more than 1,000 village names which he regarded as of this type. Many of the supposed *-ing* names were wrongly interpreted, however, and since also his now discredited theory of the 'mark' formed an integral part of his work, little attention has been paid latterly to his conclusions. It is a striking fact, however, that the vast majority of the names he lists lie within the southeastern quadrant of England—a fact consonant

---

[1] E. Ekwall, *English Place Names in -ing* (O.U.P., 1923)
[2] J. M. Kemble, *The Saxons in England* (Quaritch, 1876)

with the historical story of the settlement. Thus, Kemble deserves the credit for the first discovery of the significance of these names.

We have preferred to adopt a conservative basis for our map, and have plotted only the positions of settlements of which the

Fig. 26   An early phase in the Saxon settlement as indicated by certain place-name types.   Solid dots represent names ending in *-ing*.   Circles represent names ending in *-ingham*

name has been studied in its successive historic modifications by Ekwall. This list is definitely not complete, but it is representative.

The map, Fig. 26, shows undoubted indications of expansion of settlement from all the initial nuclear regions. Jutish settlement extends over parts of the Chalk dip-slope, west of the Stour, and also into the Lower Greensand region. Numerous South Saxon settlements have sprung up on the scarp-foot bench, and in the sandstone area, later to become the Rape of Hastings.

To a smaller extent a lodgement in the Rother valley region and the central Weald is indicated. The East Saxon settlements have multiplied in the area of the boulder-clay plateau and the more dissected country along its Londonward edge. The East Saxon group is separated by a belt of sparse settlement in southern Suffolk from the northern East Anglian area, which shows a higher settlement density than any other tract. The valley of the Thames and those of its southern tributaries show some signs of occupation, but the North Downs plateau in west Kent and Surrey, the Chiltern plateau, the 'Bagshot country', and the clay terrains north of London are almost unsettled. A beginning of scarp-foot settlement along the North Downs in Surrey is evident.

A curious feature of the map, and one somewhat difficult to explain, is its silence concerning the West Saxon expansion. That the Wessex of tradition—the western downlands—should fail to show the early type of suffix is not surprising; it merely confirms archaeological evidence. But the Icknield zone and the Oxford basin are hardly in better case, despite the clear evidence of initial settlement given by burial-grounds. There appears to be no ready solution of this enigma, but we are prepared ourselves to follow the indication of the place-name map and to conclude that, following an initial penetration of the area, its closer settlement was deferred.

It remains to point out that if we adopt a larger and looser category of names as, in the broad sense, 'early', our results can be somewhat extended. In Fig. 27 we have plotted all place-names with the suffixes *-ing*, *-ham*, and *-ton*, regarding them as an earlier group than those names which embody a topographic element, such as *-field*, *-ley*, etc. The marked difference in distribution between these two groups in Essex has previously been noted by one of us,[1] and for that region at least it is reasonable to claim that a geographical interpretation gave independent support to the idea that the two groups did in fact largely represent two independent 'crops' of names. The claim of the *-ham* group to be regarded as early is not in doubt, though there are considerable possibilities of confusion with the non-habitative suffix *-hamm*, signifying 'enclosure', 'pasture', or 'meadow'.

---

[1] S. W. Wooldridge and D. J. Smetham, *Geog. Journ.* **78** (1931), 243

The names ending in O.E. -*tun* appear to be relatively later. Mawer remarks [1] that 'the farther we move from the areas of earliest settlement, the fewer become the -*hams* and the more numerous the -*tuns*'. Nevertheless, the inclusion of a great number of the -*tuns* in a generally early group seems reasonable. In any case, the plotting of the names is worth while as an

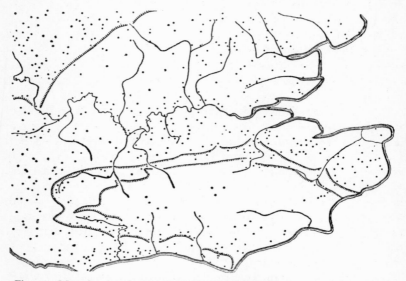

Fig. 27   Map showing the distribution of place-names ending in -*ing*, -*ham*, and -*ton*.   The lines of the Chalk and Greensand escarpments are indicated

experiment, and is justified in the result.   On any hypothesis the -*ing*, -*ham*, -*ton* distribution gives us a picture which, however inaccurate in detail, must perforce mark a further stage in the process of settlement.

A study of Fig. 27 shows that the main areas of further expansion are as follows : the Wessex Chalk area and the western portion of the South Downs, the Sussex Levels, and the Rother valley, the Thames and Lea valleys, and the vales of Oxford and Aylesbury.   The beginnings of settlement in the central Weald, the Bagshot country, the Chiltern plateau, and the Middlesex and Hertfordshire uplands are also apparent.   The

[1] A. Mawer, *Problems of Place-Name Study* (C.U.P., 1929), p. 123

picture thus presented is essentially credible and consistent with the hypothesis of a steady expansion, geographically guided. The main negative regions of the earlier phases are still clearly marked.

Finally, we may take Fig. 28, showing the distribution of the vills mentioned in the Domesday Survey, as portraying the terminal phase of the Saxon settlement. It clearly demonstrates the results of the general process of 'filling up' the countryside. Generally speaking, we should expect distribution maps of an

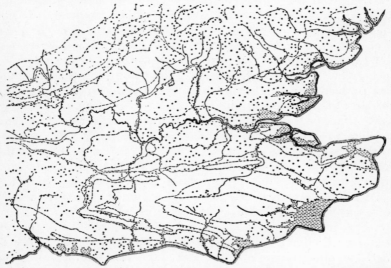

Fig. 28 Map to show the distribution of Domesday vills. The physiographic boundaries shown may be recognised by reference to Fig. 24, or a ¼-inch Geological Map

early stage of settlement to bring out the optimum or positive areas, while those of a terminal phase would normally serve to define more clearly unfavourable or negative areas. Such is certainly the case in the example before us. We note a consolidation of settlement in all the early nuclear regions, together with the results of the later long-continued secondary settlement. Thus in east Kent the Lower Greensand belt is now fully settled, while a string of settlements in the Holmesdale zone extends from the Medway to the Wey. Similarly in Sussex, the scarp-foot terrain and the Rother valley are very fully occupied, constituting, as it were, a second nuclear strip within the South Saxon region. In Essex, settlement has spread widely over the boulder-clay lands and reached a particularly high density in northwest Essex and east Hertfordshire, where the settlement is largely of a plateau character. The heavy population of the Wessex Chalk lands, particularly in their northern part, reflects another important phase of the later or secondary settlement ; it is in this region that the map contrasts most strongly with its predecessors.

Even with the virtual completion of the secondary settlement, however, considerable tracts were left thinly peopled. Taken together, these tracts form a crescent-shaped area encircling the Lower Thames basin on its northern, western, and southern sides, but leaving it open to the sea. The constituent elements of this incomplete 'barrier-ring' are : The Chiltern plateau, the Beaconsfield gravel plateau, the Bagshot country, the heights of the western Weald, and the central Wealden region. Here we have a fact of leading importance in the historical geography of southeast England, and one whose significance is not confined to the pre-Conquest period. The former barrier qualities of the central Weald have long been realised and perhaps even somewhat exaggerated, but the northern portions of this composite barrier strip have not received equal attention. They remained for long clearly traceable on the map or on the ground as distinctive elements in the physical and cultural landscapes of southeast England, and only now, in the very latest phases of metropolitan growth are they losing the last semblance of the former quality, in virtue of which they so largely controlled the settlement of the past.

In addition to the works referred to in the text, the following will be found of interest and value in studying the subject :

R. H. Hodgkin; *A History of the Anglo-Saxons*, vol. 1 (O.U.P., 1935)

J. E. A. Jolliffe, *Pre-Feudal England* (O.U.P., 1933)

T. O. Kendrick and C. F. C. Hawkes, *Archaeology in England and Wales. 1914-1931* (Methuen, 1932)

G. Sheldon, *The Transition from Roman Britain to Christian England* (Macmillan, 1932)

Reference should also be made to the publications of the English Place-name Society and to the Ordnance Survey map of 'Britain during the Dark Ages'.

13

The Loam-Terrains of Southeast England and their Relation to its Early History

IN HIS most valuable and stimulating work on *The Personality of Britain*, Dr Cyril Fox has assembled a large body of archaeological data, and has presented an interpretation of the several successive distributions based upon essentially geographical considerations. This theme covers a wide range of topics which interest workers in other fields, none more than geographers. One of the geographer's main interests lies in the study of regions, and he may well claim to have been afforded by Dr Fox a most valuable addition to his data, which enables him to push back his study of regional distinctness into earlier periods than those with which history deals. In return, he may hope to contribute something to the interpretation from his own field which embraces analysis of regional physique in all its aspects.

It is from the latter point of view that the following brief contribution is offered. It is concerned with the physique of southeast England in its relation to the earlier stages of settlement. The authors approach the subject in its archaeological contacts with some diffidence, and would wish to make clear that their aim is not to criticise but to supplement Dr Fox's interpretation so far as it concerns southeast England. It is evident that generalisation based upon the whole or the larger part of British area must apply with varying emphasis to its several parts. The primary line of division of the country recognised by Fox is the ' Exe-Tees' line of the geographer. The area lying southeast of this line certainly constitutes a major unit, when judged in respect of position and relief, but it may fairly be contended that southeast England, i.e. the area included within the main chalk escarpment as far north as the Wash, possesses certain features, especially soil features, which mark it off from the rest of the English Plain and which have powerfully affected its reactions in the history of settlement. Chief amongst these soil

223

features is the existence of extensive tracts of loamy or intermediate soils, which present optimum conditions for cultivation, but which are not explicitly recognised by Fox and other workers who have dealt with the region.

We may briefly restate Fox's main thesis in respect of population and settlement, as a basis of discussion. He draws a fundamental distinction between the porous soils derived from chalk, sand, and gravel and the heavy clay soils. In this way he is led to recognise 'areas of primary settlement' on soils of the first type, and 'areas of secondary settlement' on the heavy clays. Using the map showing massed Bronze Age finds as an index of population for that date, he calls attention to the importance of the upland chalk and limestone areas, and suggests that it was the *extent* of the areas of porous soils which determined their selection for occupation. He comments further upon the mass of finds along the Thames valley, as affording an instance of an area of a different kind in which the relatively small extent of porous soils was probably compensated for by the attractions of transport, trade, and fishing. He also notes the local concentrations of evidence near what were in essence bridge-port sites, where medieval towns subsequently sprang up.

Fox further comments on the fact that, though in the Beaker period of the early Bronze Age the areas of porous soils were naturally far from being completely taken up, the expansion and increase of population during the ensuing thousand years still left considerable tracts of such porous soils unoccupied. In explanation he suggests that before the expansion was complete, increasing skill in agricultural methods permitted the beginning of exploitation of the clay lands, the areas of secondary settlement, and he regards such secondary settlement as conditional upon, not only a higher standard of agriculture, but of civilisation as a whole, in periods of greater political security. Chief amongst such periods were those of the Roman occupation and the later phase of the Anglo-Saxon expansion. It is to the second period that he would attribute the main phase of attack on the forest lands, regarding the smaller beginnings evidenced in Iron Age and Roman-British times as in some sense 'false starts' in a general process. He notes that the existence of the pre-Roman Belgic dynasty was a not unimportant phase in the process which led to a 'shift in the economic centre of Britain from the Chalk plateaux

—the Salisbury Plain region—to the richer lowlands of East Anglia'. In spite of this change it is pointed out that London had not become in any sense a regional focus during Iron Age or earlier times—but came into being only in the Roman phase, when penetration of the surrounding woodlands first became a really practicable possibility. Dr Fox suggests that the general movement to the areas of secondary settlement is what is really implied in the oft-used term 'valley-ward' movement which was a change in mode of exploitation rather than in habitation—a move from the poorer to the richer soils.

With most points in the general thesis thus elaborated by Fox, the authors are in cordial agreement. The main ground of difference may be summarised in the suggestion that it is not sufficient to divide the soils of the region into two classes, essentially permeable and impermeable. It is, of course, evident that there are many local variations which cannot in any case figure in a general survey ; that this is the case Fox clearly states. Our point, however, is that a third main group of soils exists, intermediate between the extreme classes. The significance of these loamy or intermediate soils in relation to settlement and agriculture in all ages cannot be too strongly insisted on. Their characters are as definite and as worthy of separate recognition as those of the chalk and sand, and clay soils treated by Dr Fox. By styling them 'intermediate' we do not simply imply that they are gradational, comprising in various degrees the characters of the extremes between which they fall. That to some extent they partake of this nature is not to be denied, but a truer view is that which regards them as a natural soil group, giving rise to highly distinctive and clearly bounded regions, which figure among the more important settlement areas of the country from Bronze Age times onward and which remain clearly recognisable in the present-day 'cultural landscape'.

The rather general failure to recognise the importance of this soil-type in Britain is attributable in part to confusion and accidents in geological nomenclature, and to the reluctance of British geologists to use the term loess for what are undoubtedly the analogues or equivalents of that well-known accumulation. The significance of the loess and related limon soils of the continent of Europe has been fully recognised by archaeologists and geographers. It may be recalled that Vidal de la Blache, in

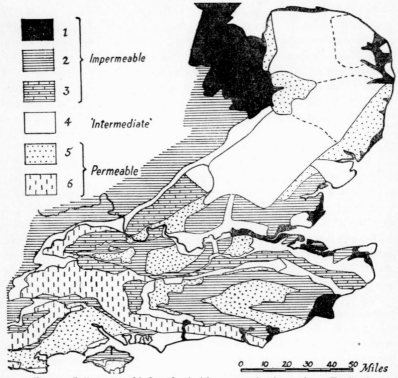

Fig. 29 Soil types and inferred primitive vegetation in southeast England

1. Alluvial soils : marsh, water meadow, and alder brakes. 2. Heavy clay soils : damp oak forest. 3. Residual and drift clays of high Chalk plateaux : dry oak wood, etc. 4. 'Intermediate' or loamy soils derived from the following parent materials : High and low-level brickearths, and certain terrace gravels ; Calcareous boulder-clay ; Thanet Marls ; Lower Chalk ; Upper Greensand (malmstone) ; Sandgate, Bargate, and Hythe beds ; original vegetation unknown, probably light woodland or scrub ; arable from early times. 5. Sand and gravel soils : oak—birch—heath association. 6. Chalk soils : open chalk grassland and scrub. The dotted lines in East Anglia delimit respectively the Norwich loam-region, and the colder, heavier boulder-clays of Suffolk

his well-known 'Carte pour servir à l'histoire de l'occupation du Sol' [1] figures the areas of loess and limon soils clearly.   In the accompanying text he shows that these soils, giving optimum conditions for agriculture, have been salient features in the complex of European conditions in both prehistoric and historic

[1] E. Lavisse, 'Tableau de la Géographie de la France', *Histoire de France* (Hachette, 1911)

times. They extend in two zones across Europe, one passing from Moravia and the Danube valley to Alsace, while the other extends from northern Bohemia via Saxony to the Hesbaye and Picardy regions. The map suggests the extension of similar soils into southeast Britain—but the areas are vaguely placed and ill defined. Indeed, little attempt has been made to recognise the equivalents of the limon soils in England and we may briefly review their distribution here.

In the first place it must be noted that extensive tracts of low-level or valley brickearth exist within the area, and that large parts of the valley floors and terrace surfaces are not so much gravel as loam-covered. The close resemblance of these deposits to loess has often been commented upon, and if this similarity be granted the question of origin need not here concern us. Among the important valley brickearth areas, we note the northern side of the Thames valley near London, especially the surface of the 'Taplow terrace'. This broad loam-covered ledge extends eastwards into Essex and also westwards from London, where it expands in southwest Middlesex into a considerable region. Reference to the newer geological maps will show how much of the area is thickly covered with brick-earth ; but it should be noted that extensions of the latter, inconsiderable geologically but none the less important in soil control, cover much of the area necessarily mapped as gravel, both here and in other localities. The brickearth areas south of the Thames are less extensive, but the clearly bounded plateau region behind Southend is of precisely similar type, and a broad tract of loam-covered terrace gravels extends up the western side of the Lea valley to the neighbourhood of Hoddesdon. Extending our examination over a wider area, we note that the seaward and larger portion of the Sussex coastal plain is a precisely similar loam-region, while similar soil conditions recur over a wide area between Norwich and the sea [1] and in the Tendring Hundred of north Essex. Less considerable tracts of lowland loams which are nevertheless deserving of mention, occur in the Medway valley between Maidstone and Tonbridge and in the Sitting-bourne area of north Kent.

Exactly similar soil conditions recur over the outcrops of the

---

[1] P. M. Roxby, 'East Anglia' in *Great Britain* (C.U.P., 1930)

high-level brickearths which occur locally on the Chalk plateaux. These are certainly of different origin from the valley brickearths and in the opinion of the authors are largely, if not entirely, true loess. They occur in scattered patches on the North Downs and the Chiltern Hills, generally overlying the true clay-with-flints, but it is only at the eastern end of the Chiltern plateau, between the valleys of the Mimram and the Beane, that brick-earth soils are sufficiently dominant to constitute a loam-region. There is some tendency for the loamy variant of the plateau drift to increase in importance eastwards along the North Downs also, and it is possible that the block of Chalk country between the Stour valley and the sea may merit recognition as a separate soil region ; but further investigation is required on this point.

So far we have been dealing with soil regions which all would recognise as analogous to the limon regions of the Con-tinent, though identity has been masked by discrepant nomen-clature. It remains to observe that closely similar loamy soils have been developed also upon several other important formations in southeast England. The realisation of this fact depends upon the examination of the soils themselves or of mechanical analyses of them ; the geological map with its relatively simple age nomenclature of itself yields little guide.

There are four chief regions or types of region in which loamy conditions occur beyond the boundaries of the brickearth tracts. In the first place we have the soils derived from the so-called marls or 'tuffeau' of the Thanet beds of east Kent. From the coast to the Medway valley and to a smaller extent west-wards to the outskirts of London, these beds yield loamy soils, the mechanical analyses of which compare closely with those of the brickearth soils. The region of optimum soil conditions thus defined is of the highest significance in the early history of the region. Soils, perhaps generally sandier but retaining the essential character and ease of working of loams, occur in the Lower Greensand region of the Weald, on the Hythe beds *east* of the Medway (which contrast very markedly with their more sandy equivalents to the west), on the Bargate and Loamy Folkestone beds of the Guildford and Godalming area, and on parts of the Sandgate beds of Sussex. Soils, again of the same type, though tending to show a higher clay fraction, are char-acteristic of large parts of the boulder-clay surface of East Anglia,

Hertfordshire, and Essex, where, moreover, they are often associated with tracts of true brickearth. In regard to the boulder-clay soils, in general, certain distinctions must be made. In Suffolk, where the drift contains an appreciable element of Kimmeridge Clay, and also along the southern margins of the great sheet, in Essex and Hertfordshire, where the London Clay has been laid largely under contribution, the soils locally approximate to the true heavy clay type. Over the remaining and larger part of the glacial area, this is not the case. While local variations of course occur, the soils are predominantly loamy in the sense here defined, and they are comparable in many essential respects with the other loamy soils we have noted. Finally we would wish to bring into the same general group of loamy soils, those which are developed upon the lower argillaceous portion of the Chalk formation. They extend over tracts of varying width at the foot of the Chalk escarpment, and the soils of the Upper Greensand bench, where this bench forms a notable feature, as in Sussex and Hampshire, are of not dissimilar type and may be grouped with them. In all cases the soils contain much more nearly equal proportions of the mineral grades, sand, silt, and clay, than do the soils derived from the purer Upper and Middle Chalks. Their ease of working is attributable in no small degree to their lime content, and to this extent they differ from the other loamy soils with which we are grouping them, but from the standpoint of human utilisation their affinity is undoubtedly with this group.

Space does not permit us to discuss in detail the other soil groups of southeast England, but we may briefly note the main types of soil or vegetation regions which occur in juxtaposition with the loam terrains and thus build the complex soil mosaic that was presented to early man. The heavy clay areas are much more restricted than is commonly supposed. They are located on the London Clay in Middlesex, Essex, and Surrey, though in the last named area and farther west an appreciable lightening element of sand and silt occurs in the soils. Their greatest expanse is on the Weald Clay, where again there is a considerable local admixture of sand and areas of drift-covering. The Wadhurst Clay of the Rother valley provides another tract of heavy clay land. Contrasted with these regions, we have the heathland tracts on the coarser sands and sandy gravels, notably parts of the Lower Greensand, the Bagshot beds, the Blackheath beds, the glacial gravels, Pliocene Crags, and the blown sands of the Breckland. On the Chalk areas we must distinguish the thin red and black soils of the true downland areas from the heavily wooded and intractable areas of clay-with-flints. Finally we have the riverine and marine alluvium which has figured as a productive soil only late in the economic development of the region, having been largely marsh or water-covered during the earlier phases.

With the foregoing brief summary of the soil conditions of the area as a basis, we may attempt to justify the claim that it was the loam-regions which figured most prominently as the nuclei of settlement and penetration during the earlier stages of peopling the country. Their pre-eminence in the agriculture of today as regions favouring arable farming cannot be questioned. We need to inquire rather at what date their favourable qualities were first recognised and turned to account. From this point of view Fox's map of 'massed' Bronze Age antiquities provides interesting data. Accepting his reasonable assumption that the frequency of finds may be taken as an index of population, we note among the densely settled tracts the loam-region of the Sussex Levels, the 'Bargate' region in the neighbourhood of Guildford and Godalming, the Southend Loam plateau—a significant coastal 'landing stage', and the Norwich region. There is also the conspicuous tract of high find density which follows the Thames valley and the lower parts of the valleys of the Brent, Lea, Wandle, Medway, and Stour. As we have noted, Fox advances good economic reasons for the occupation of the Thames valley floor; but to these we would desire to add the important fact that the broad loam-covered terraces offered opportunities for agriculture not essentially inferior to those of the chalk lands. Of the original vegetation cover of the brickearths we know little or nothing; but it is a reasonable surmise that they carried relatively light and easily cleared woodland diversified by some heathland country where the underlying gravels emerged. Our point, essentially, is that the Thames-side tract and its analogues in the tributary valleys were not merely regions of porous soils rendered tolerable by the attractions of trade, transit, and fishing. Their extent, it is true, was less than that of the still favoured downlands, but otherwise they presented optimum conditions for clearing, cultivation and, be it added, for obtaining an adequate water-supply.

Certain other features of the Bronze Age map are worthy of note. Though the major congregations of evidence outside the Thames valley are on the Chalk of Wessex and in the Breckland and contiguous tracts, among the latter we note what we may call the Icknield Zone (as far as the Hitchin Gap) which *inter alia* shares the loamy soil characters of the valley tracts. Moreover, the Essex boulder-clay lands show evidence of considerable occupation. The most notable 'negative' areas, where evidence is absent or scanty, are the larger part of the Chiltern plateau, the London Clay areas of Middlesex, Hertfordshire, and Essex, the sandy Bagshot plateau, the New Forest, the Weald Clay terrain west of the Arun, and the larger part of the Weald of Kent. These facts appear to indicate that the sandy areas—the areas of porous soils *par excellence*, were not sought as such. Some of the Wealden heathlands were, as Fox remarks, 'islanded in forest' and thus inaccessible—but not all of them were so placed. The Bagshot plateau abutted upon the Thames valley with no considerable intervening forest tract. It appears to be reasonable to claim that the truly sandy areas and especially those of upland or plateau character never attracted occupation during early times except during the curious and interesting Mesolithic phase recently studied by Clark.[1] It seems to us that with certain exceptions, such as the Breckland and the coastal Suffolk heaths, which are relatively low-lying and which enjoyed exceptional advantages of access, the sandy tracts were repellent rather than attractive. If, however, Fox is right in supposing that part of them at least remained unoccupied because of superior attractions of other areas in a new agricultural phase, then we may maintain that to a large extent it was to the loam-terrains and not to the clay lands that population moved.

We may test the significance of the loam-areas further by reference to Iron Age distributions. The distribution of Hallstatt-La Tène I–II pottery in the area does not bear decisively on the point at issue, for the finds are relatively sparsely distributed and demonstrate little more than the continued occupation of certain of the chalk lands and the importance of the Thames

[1] *The Mesolithic Age in Britain* (C.U.P., 1932)

valley and the Icknield zone. But with the Belgic phase, recently so thoroughly studied by Hawkes and Dunning,[1] a number of relevant points come to light. The general distribution of pedestal-urns (excluding those of Roman date) immediately suggests the importance of (*a*) the Kent Lower Tertiary zone and (*b*) the Essex and Hertfordshire boulder-clay lands. All the evidence combines to suggest that the earlier wave of Belgic peoples, entering by the Kentish angle, settled in great numbers on the belt of rich arable soils situated upon the brickearths and Thanet loams, and on the lower portions of the Chalk dip-slope where considerable relics of these formations survive and modify the character of the soil. There is no reason indeed to suppose that they were confined to this zone ; penetration through the Medway Gap is clearly indicated. Nevertheless it cannot be supposed that the upper part of the Chalk dip-slope with its clay-with-flints cover was other than densely forested and this forest formed a natural boundary to the loam region. The oft-quoted observation of Caesar concerning the density of population and the ready availability of corn along his line of march is relevant here—and it is perhaps not too much to suggest that the line of Watling Street itself depended as much upon coincidence with a relatively narrow zone of favourable soils as on the possibility of bridge crossings.

As regards the slightly later phase of the Catuvellaunian dynasty, centred at St Albans and afterwards at Colchester, it is clear that we are dealing with a drift-land, not a clay-land distribution. Hawkes and Dunning extend the limits of the Catuvellaunian territory southwards to the Thames, northwards into Northants, and on the evidence of coins, westwards to the Cherwell valley. That their 'sphere of influence' extended over such an area is not to be gainsaid on the evidence, but that the whole of this tract showed even approximately equal density of settlement is inherently improbable. The main mass of the loamy boulder-clay country terminates at a line drawn from Hertford to Hitchin. The area favourable to cultivation may have extended somewhat west of this over the loam-covered eastern end of the Chiltern plateau, and in any case it must have been prolonged for some distance westward along the line of

[1] *Archaeological Journ.* 87 (1930)

the Vale of St Albans, which was entered by the ice-sheet at its northeastern end. The most westerly record of a pedestal-urn from this district is from Abbots Langley on the northern margin of the Vale. To the south of the driftland tract stretched the London Clay forest, which was but little penetrated or settled even in Roman times. The rarity of pedestal-urns of pre-Claudian date in the Thames valley near London, suggests that the pre-Roman Belgic settlement did not extend appreciably into the Thames valley in spite of the break in the forest barrier offered by the Lea valley. On the north and west of the settled driftland area lay the accidented and heavily forested tract of the Chiltern plateau, which retained in some sense its 'barrier' quality till Saxon times. It is in the light of these facts that we claim that the heart of the Belgic kingdom lay upon the boulder-clay lands, and that westward penetration was probably largely confined to the region of the Vale of St Albans and the south-westerly continuation of the 'Icknield zone' beyond Hitchin. In this connection again, we would venture to suggest a slight modification of the attractive thesis put forward by Fox. He comments upon the shift of the economic centre of Britain at this date from Wessex to East Anglia, and instances this as a phase in the taking over of the clay lands. But as we have seen, the region in question is not a true clay land nor does it extend into East Anglia. To the north of the lighter Essex boulder-clay lies the heavier Kimmeridgic variant of Suffolk, a region much less tractable under axe and plough, and no doubt serving as a natural barrier against the earlier established Celtic Iceni. It was to Essex and east Hertfordshire—a true natural region of ready clearance and cultivation, naturally invested by forest,— that the economic centre moved ; and the accessibility of the southern boundary of this region to the crossing place of the Thames near London has a clear bearing on the Roman initiation of the latter city. The historical significance of the Essex drift lands in Roman times is briefly discussed elsewhere,[1] in collaboration with Mr D. J. Smetham, and we shall not here pursue the theme. We may recall, however, the important geographical fact that the loam-areas settled by the Belgae in Kent and Essex respectively and inherited by the Romans, converged

[1] *Geog. Journ.* 78 (1931), 243 ; and Essay No. 11, p. 171 of the present volume

upon the London crossing from the coast and together constitute one of the major significant elements in the siting of London.[1]

To what extent the true clay lands were cleared during Roman-British times it is difficult to say. We here touch the fringes of the interesting topic recently discussed by Collingwood,[2] Randall,[3] and Wheeler [4]—viz. the general nature of the agriculture and population of Roman Britain. Collingwood has argued that the total population of this phase was small and that its distribution was of upland or 'prehistoric' type. He supposes that agriculture was still in essence primitive—not greatly in advance of subsistence farming—and that there was a general reluctance to incur the capital expenditure involved in the clearance of low-lying forested country. A number of considerations have been urged against these views—and among the most cogent is that based upon the well-attested exports of corn about A.D. 360. On the whole it appears to us that Collingwood's case for an essentially primitive agriculture is difficult to maintain, but in the discussion which his ideas evoked, the facts of distribution as shown on the Ordnance Survey map of Roman Britain, from which he started as a basis, have been rather lost sight of. It therefore seems worth recalling that whatever the productivity of agriculture during this phase, and whatever improvements were introduced by the Romans, the facts at present known do not suggest any considerable attack on the true forested clay lands. The main 'negative' areas on the Roman map are quite clearly and definitely as follows : the greater part of the Chiltern and Bagshot plateaux, the London Clay areas of Middlesex and Essex, the central Weald and the heavier boulder-clay lands

---

[1] It may be noted that 'Belgic Frontier' defined by Wheeler (*Antiquity* 6 (1932), 133ff.) coincides remarkably closely with one of the soil-boundaries shown on the accompanying map (Fig. 29) ; to the north lies the Chiltern plateau, while to the south is the gravel and loam terrain of the Vale of St Albans. The tract of 'pre-historic forest' shown on Wheeler's map is presumably based upon the distribution of the boulder-clay as shown on the geological map. In actual fact the boulder-clay is partly replaced by and almost completely covered by brickearth in this area, which must thus have been favourable to arable cultivation. . The heavy forest land probably came on suddenly north of the ' frontier', which is a natural ecological boundary.

[2] *Antiquity* 3 (1929), 261   [3] *Antiquity* 4 (1930), 80
[4] *Ibid.* p. 91

lying north of Stane Street in Essex. This distribution is virtually identical with that for earlier periods and signifies an avoidance of clay country, and of tracts of sandy upland. There is little doubt in our view, however, that the farmed country was very much more productive than Collingwood supposes, for to the facts adduced by Randall we may add this salient geographical consideration—that the lands farmed by the Romans in Essex, Hertfordshire, and Kent are among the most naturally productive in the region and are still indeed notably preponderant as areas of arable farming. The soils of these regions belonged neither to the lighter nor heavier classes as commonly defined, but to the vitally important intermediate group we have sought to distinguish. Herein lies a reasonable explanation of the combination of high productivity with the general absence of areal expansion.

Into the long and obscure history of the Saxon penetration and settlement of the area we must forbear to enter here. It presents innumerable points of interest in connection with soil geography, and we hope to treat of these elsewhere. For present purposes it must suffice to note that the later part of the Saxon period appears to have witnessed the first real inroad on the clay forest lands. Such indeed is Fox's conclusion and we may urge in further support the evidence of place-names. Those believed to be of later date certainly congregate markedly on some of the clay lands, as in southern Essex.[1] On the other hand the place-name evidence as a whole does not lend support to the idea that expansion into the woodlands was delayed altogether until after the pagan phase. Fox [2] interprets Leeds' map of cemeteries as indicating that hardly any progress was made in pagan times, but Leeds himself, contrasting the narrow localisation of the cemeteries in East Anglia with wide distribution of Anglo-Saxon names (by no means wholly of later type), was led to remark that 'it demonstrates how soon the inhabitants must have ceased to be influenced by this former selectiveness' [3]. At any rate we may reasonably conclude that the surviving

[1] Wooldridge and Smetham, *Geog. Journ.* 78 (1931), 260
[2] Sir Cyril Fox, *The Personality of Britain* (Nat. Mus. of Wales, 1947), p. 71
[3] E. T. Leeds, *The Archaeology of the Anglo-Saxon Settlements.* (O.U.P., 1913), p. 69

tracts of boulder-clay woodland—if not the clay forests farther south—were appreciably cleared in relatively early stages of the Anglo-Saxon settlement. In the general picture of Saxon times the loam-terrains figure as the regional nuclei of the several kingdoms, much as they did during the late Celtic phase, and we may note further that in this stage we witness the first considerable settlement of the loamy Lower Greensand tracts in east Kent and west Sussex, which become, as it were, secondary nuclei of the Kentish and South Saxon kingdoms, pendent respectively upon the Lower Tertiary loam belt and the 'Sussex Levels'.

In conclusion we may remark that the theme here briefly presented is capable of, and indeed calls for, a more extended treatment which we hope to devote to it in due course. What we have termed the loam-terrains constitute an essential key to the study of the later phases in the historical geography of the area, and even in the present phase of 'metropolitan' geography they remain agriculturally distinct. In the foregoing paragraphs we have sought to trace their significance in the archaeological field only, and we submit that they stand out as areas of 'primary settlement' first entered upon in the Bronze Age, more extensively taken up in the Belgic phase of late Iron Age times, inherited with but little extension or areal modification by the Romans, and finally consolidated and extended as regional nuclei by the Nordic invaders, prior to the settlement of the latter within the areas of the heavy clays. The treatment here attempted emphasises the essential similarity in soil-character of regions which, judged by examination of a geological map, would appear entirely dissimilar. If the group of loamy soils is included within the major division of Fox's porous soils, the main lines of his generalisations appear to be abundantly supported by the facts. As we have indicated, the rather loose colloquial soil terminology, combined with certain imperfections in geological nomenclature, are apt to conceal the existence of the loamy-soil group, placing some of its members with the true clays, to which they bear little resemblance and linking others with the coarser sands and gravels, from which again the differences are radical. If independent status is given to this soil-group, the study of the progress of early settlement is considerably clarified. The very partially true connotation of the term 'valleyward movement' commented upon by Fox and by Wheeler [1] is also rendered clearer. In southeast England it was to plains and low plateaux of a definite soil constitution as well as to valley floors that Early Man progressively moved.

[1] *Antiquity* 4 (1930), 92

14

Some Geographical Aspects of the Greater London Regional Plan

1 *The Major Features of the Plan*

DESPITE the complexity of its matter and the magnitude of the problem it essays to solve, the main outlines of the Greater London Regional Plan are clear, simple, and readily comprehended. It assumes, following the Barlow report, 'that no new industry shall be admitted to London and the Home Counties except in special cases', and further that the total population of the area will not increase. Another assumption which becomes, in fact, the virtual mainspring of the actual plan, is the need to decentralise or 'decant' population and industry from the congested central areas of the County of London and the adjacent zone of 'Greater London'. With a recommended maximum density in these central areas of 136 persons to the acre, it becomes necessary to plan for an exodus of roughly 600,000 from the County, and a further 400,000 from beyond its boundary. The essential aim of the Plan thus resolves itself into a 're-arrangement of population and industry within the region'.

The Plan then proceeds to define and describe the existing faintly marked structure of the urban mass in terms of concentric zones. It recognises a discontinuous Inner Urban Ring, comprising, as it were, the overspill of the closely built up areas of the County. Beyond lies a broad and ragged fringe, the Suburban Ring, with lower building density. The Plan then defines with some care the outer limit of the Green Belt Ring, which includes the greater part of the land acquired under the Act of 1938, and considerable stretches of open country, together with a number of separate or island communities, some of them old towns and villages, and others 'upstart' dormitory suburbs. Beyond lies the Outer Country Ring, comprising the rest of the planned area.

It is the avowed aim of the Plan to adopt and to define more strictly this rough zonal scheme, imposing 'directive and corrective aims'. The Suburban Ring is treated as essentially static, in the sense that it neither requires decentralisation nor is it fitted to serve as a reception area. For the Green Belt Ring it is proposed that there should be stringent and permanent safeguards against building, so as to conserve existing stretches of open country in private ownership, as well as to facilitate the long-term realisation of the original Green Belt concept—the acquiring of 'recreational' land by public authorities. A certain, though strictly limited, growth of existing communities must perforce be allowed, but it is proposed that no new centres be established. The logical end of the policy outlined thus casts the Outer Country Ring for the rôle of chief reception area for overcrowded London. To this end, the Plan proposes considerable though controlled growth in existing towns within the Ring, and also the planting of new satellite towns. This last feature is one of the most distinctive and at the same time most debatable elements in the Plan. Ten sites are proposed, at White Waltham, Redbourn, Stevenage, Stapleford, Harlow, Ongar, Margaretting, Meopham, Crowhurst, and Holmwood. Assigned a maximum population of 60,000, eight such towns would house a considerable part—approximately a third—of the displaced central population and industry. Of the rest, roughly a quarter of a million are allocated to existing towns in the Outer Country Ring, and a similar number to points outside the region. This leaves 125,000 for disposal in what are termed 'quasi-satellites' in the Green Belt Ring.

The foregoing brief summary necessarily does less than justice to the Plan as a piece of detailed and coherent thinking, but it serves as a necessary introduction to the present paper. We are not here concerned with the structural details of the Plan, its proposals for roads, railways, and the like. Such arise as necessary ancillaries of this or any other Plan.

The broader aspects of the Plan will necessarily evoke comment and criticism from geographers, economists, and others, and since it is *par excellence* a 'social document' of immediate concern to millions, it is entirely right and proper that such comment and criticism should find place. But here a distinction must be made. It is evidently possible to criticise the Plan in the light of large political and economic considerations which bear upon its basic assumption and declared general aims. Some will lament the passing of a certain sort of so-called 'freedom', and others will declare that the Plan is, in the narrow sense, 'uneconomic'. To still others it may seem, with Professor Dennison, that 'we do not know the shape of the post-war economic world', and that this ignorance fatally prejudices any plan. All that need be said here is that none of these considerations is in the mind of the present writer. The following account recognises the need for such a Plan as the one before us, accepts its assumptions, and concurs in its major proposals. It is to be read, therefore, not as an attack upon the Plan, but as a critical commentary upon it by one whose field of study for many years has embraced the geology and geography of the Metropolis and its countryside.

2 The Physique and Extent of the Planning Region

A feature which cannot fail at once to impress the geographical reader of the Plan is its virtual silence on the physical constitution of the region dealt with. In one context only, that of agriculture, is the question of land character raised, and Professor Stamp's chapter, while fulfilling its purpose of raising the agricultural issue, is fitted neither by its contents nor by the position accorded to it in the volume, to serve as an introduction to the whole. It would not be fair, indeed, to claim that the Plan entirely ignores 'physique'. It acknowledges the existence, and proposes to conserve the amenities, of certain major features, such as the North Downs scarp, the Leith Hill range, and the lower Lea valley. It remains true, however, that there is no preliminary 'regional survey' providing frame and background to the immediately practical problems. The Plan (Par. 42) avers that 'the London Region surveyed by the impartial and comparative eye of the geographer cannot be said to be very remarkable for the possession of dramatic, romantic, or noble landscape features'. To this we may reply that the said 'impartial and comparative eye' sees, or should see, much more than the dramatic, romantic, and noble. 'If the culture of the environment,' writes Mumford,[1] 'had yet entered deeply into our consciousness . . . we should have an equal regard for every nook and corner of the earth

[1] *The Culture of Cities* (Secker, 1940), p. 332

and we should not be indifferent to the fate of less romantic areas.' In such terms the geographer or naturalist familiar with the London district is certainly likely to react to the planning of his natural laboratory, nor is the 'visual solace' of the layman likely to be well served on any less exacting terms. In short, planning and conservation demand thorough preliminary stock-taking of resources, both economic and aesthetic. Without such stocktaking it inevitably limps on one leg and its proposals are subtly falsified.

It is, on the other hand, perhaps not difficult to see certain compelling practical reasons for the suppression, in this case, of the usual preliminary 'regional survey'. The Plan is, in a sense, an emergency measure, seizing the occasion of a major pause, both psychological and sociological, in the growth of London. As such, it had a case for preferring speed to leisurely width of view. Again, it might be urged that the physique of the London region is, in fact, well known and adequately presented in excellent topographical and geological maps. How far this statement is really justified we shall consider below. A more cogent argument might seem to lie in the fact that the latter-day growth of London has been almost independent of physical controls. The outer limit of the Suburban Ring consistently remains close to the twelve-mile circle, centred at Charing Cross. The growth stages it represents proceeded almost unhindered, across hills and valleys, clay sand, and gravel alike. We are the witnesses of a complete and almost unparalleled transformation of an area of the Earth's surface, and one which obscures the natural environment to vanishing point. It might, indeed, appear a fair claim that if London had grown in like fashion to its present form on a featureless plain, athwart a symmetrical estuary, it might have offered similar, if not identical, problems to the planner. Such a metropolis, like the London of reality, would have constituted in itself a fact of geography and might have confronted us with severe internal congestion, combined with 'marginal sprawl'. With similar governing assumptions, a similar Plan might have emerged, proposing to decant the central surplus of population and industry to an outer fringe of satellites or elsewhere.

It would seem that, for better or worse, views such as this have tacitly governed the Plan in large part. It is too dominantly architectural, in the wide sense, and too little geographical. We must not, indeed, underestimate the 'determining effects of existing conditions' which, however unwelcome, are amongst the 'given' limitations on the planner. One seems to hear an echo of a much quoted statement once made by Mackinder [1] :

> To a person about to be operated on for appendicitis, the history of the development of the appendix and of its degradation from an old function is not, as a rule, of much interest. The London of today is a vast organ in the anatomy and physiology of the Earth's surface, a deposit of several million human beings inter-related in the most vital and complex ways with one another and with the soil on which they are placed. The Geographer has to appreciate and weigh that great fact. The characteristics of the little hills and marshes and bits of forest upon and amid which the original village or villages were placed . . . are a totally different matter.

But this statement, in its context, was designed to illumine the distinction between the geography of the present and historical geography ; the antithesis which it presents is not one which in any way concerns us here. The statement, coming as it does from the pen of a great geographer, includes, as we should expect, the significant words 'and with the soil on which they rest'. The 'inter-relations one with another' are a fit field for sociological study. A close study of the past stages of London's growth might engage the historical geographer without throwing much light on the future envisaged by the planner. But the 'soil' on which London rests is still a factor in its life and future. One uses 'soil,' of course, in its wider connotation, as covering structure, relief, hydrography, or, compendiously, 'physique'. In townplanning, properly so called, the features of the *site* may or may not be important ; often they are 'recessive'. But Greater London is not a town but a region, and it requires treatment as such. It has been said that 'the re-animation and rebuilding of regions as deliberate works of collective art is the grand task of politics for the opening generation'.[2] What, then, is the London 'region' on which we are to work ?

The Restriction of the Planning Area and its Effects. Any attempt to answer this question brings us forthwith to the realisation that the area treated in the Plan is too small. Whatever considera-

---

[1] *Geog. Journ.* **78** (1931), 268   [2] Mumford, *op. cit.*, p. 348

tions of time, expediency, or economy informed the limitation of area, it flies in the face of the geographical facts and might well doom the Plan to disastrous failure. The London region, properly so called, while it is necessarily less than the Metropolitan England of Mackinder, is at least very much more than Greater London and its immediate environs. To a very close approximation it may be regarded as equivalent to the area styled by Fawcett[1] the 'London Province'. Although the detailed demographic basis of Fawcett's divisions is now a quarter of a century out of date, this in no way diminishes the force of his arguments, nor, in the case which here concerns us, is the geographical validity of his suggested boundary at all easily impugned. The London region or province must be deemed to include the whole of the counties of Middlesex, Essex, Surrey, and Hertfordshire, together with Buckinghamshire south of the Chiltern edge, and Kent, Surrey, and Sussex. It comprises, in fact, the whole of the London Basin east of the meridian of Maidenhead, together with the Weald. Providing that it is justifiable to recognise separately an 'Oxford-Reading' or 'middle and upper Thames' region there is no serious fault to find with the Plan boundary west of London, but elsewhere it is without geographical meaning. It excludes the Chiltern country beyond High Wycombe, the larger part of both Essex and Kent, and the whole of Sussex. The disruption of the Wealden area can be justified on no grounds whatsoever, aside from administrative convenience. The urban population of the south coast is in fact as well as in name 'London by the Sea'. Kent coal, while unable to compete on Thames-side with sea-borne coal from farther afield, is an undoubted integrating factor within the region. Nor must we lose sight of the possibility that further resources of coal or petroleum may be discovered in the area. But the implications of the matter are most clearly seen in the Electricity Grid, which, as early as 1927, formed an integrated network, with generating stations at Hastings, Tunbridge Wells, Maidstone, Gravesend, Barking, Croydon, Guildford, and Brighton. With the Weald, indeed, it is surely a question of all or none, and since London now directly encroaches on its northern border, there is, in fact, no possible doubt that the whole of the Weald should be included in any plan. To

---

[1] *Provinces of England* (Williams and Norgate, 1919)

leave Sussex as an area for independent planning, lying as it does, in relation to the economic life of the country as a whole, in the lee of London, is to condemn ourselves to complete frustration and chaos in planning.

The effective northern border of the London region might seem more open to question, and if we limit ourselves by the assumption that the metropolitan functions in such regions must be concentrated in one city, the difficulty must remain unresolved. Neither central England nor East Anglia as defined by Fawcett shows a single completely dominant metropolitan focus. It can hardly be doubted, however, that we must recognise, sooner or later, the principle of 'polynucleation' in the organisation of regions, the sharing of metropolitan functions among towns of lesser rank. Only so can we meet the objection that we do not pass northwards out of the London region until we enter the Birmingham region.

On grounds such as these alone it cannot, therefore, be doubted that the area of the Plan is too small, a fact, which is further indicated by the need found to accommodate part of the displaced central population at points beyond the planning boundary. This immediately raises the further question of the effect of the Plan, if carried out, on those parts of the London region beyond the adopted boundary. We were confronted before the recent war by a powerful tendency for population and certain forms of industry to move into or towards Greater London. The chief active stimuli were the pull of the great urban market and the minimisation of costs in export trades produced by proximity to the London docks. If the area of the Plan alone were subject to restrictions on the entry of industry, resumption of centripetal shift would simply lead to the massing of industry and population immediately beyond the boundary of the forbidden ground. Migrating industry could still move to points within easy road-haul of tide-water, and the result could be none other than the continued urbanisation and industrialisation of the outer parts of the London region. The strategic and social consequences of this would be little better than those of a continued growth of London itself. The difficulty involved manifestly illustrates the incurable defects of a system of piecemeal planning working outwards from a metropolitan centre. The County of London Plan involved commitments carried over into the

Greater London Plan, and this in its turn could not be implemented without definite repercussions on the country beyond its borders. There is evidently need of a complementary 'London Region Plan', to say nothing of a National Plan.

We must note further that the limitation of area has a constricting influence on the Plan itself. From some points of view, at least, the satellites appear to be too close to London, at an average distance of twenty-three miles. The effect is to overload with urban development some sections of the zone immediately beyond the Green Belt Ring. In the country north of London an area of rather more than 300 square miles, a square of eighteen miles side, is destined to include not only the satellites of Redbourn, Stevenage, and Stapleford, but also the following towns in which a considerable measure of further growth is envisaged : St Albans, Hemel Hempstead, Harpenden, Luton, Dunstable, Hitchin, Letchworth, Welwyn Garden City, Hertford, and Ware. A point in the upper Mimram valley near Whitwell, will lie near the centre of a circle of these towns, distant some five miles. Such an urban group, if it be less than a conurbation, will certainly constitute a 'constellation', and the eastern Chiltern area will be very effectively urbanised.

Elsewhere the difficulty is less marked. The Essex satellites are notably isolated. Meopham would, however, form a group with Gravesend and the Medway towns, and the Surrey satellites a similar group with Dorking, Reigate, Oxted, and Horley, while White Waltham is close to Maidenhead, Windsor, and Slough.

There is, of course, a reasonable reply to this plaint of congestion in that the satellites need, by their very nature and purpose, effective economic and social liaison with London. They have, therefore, been placed as close to London as the exigencies of the general plan permit, and the real governing conditions are the minimum requirements for a Green Belt Ring. It does not follow, however, that with electrified rail services, the satellite rôle need be lost at considerably greater distances, and, with the regional net cast wider, a greater range of possible sites would have come under review. Moreover, it cannot be doubted that the satellites will also provide urban functions and amenities for the country in which they are set. The area limitations imposed on the Plan might well result in an over-provision of such facilities in the 'twenty-five mile zone', already within reach of London, and a corresponding starvation further afield.[1]

The Problem of Water Supply. In respect of the satellite towns there is further ground for doubt and scope for inquiry in connection with water supply. It is not as widely realised as it should be that, as contended in recent years by leading consulting geologists, the absolute population ceiling in any region is set by the availability of water. It might appear that, providing the population is re-distributed but not increased, no problem could arise. The overgrown London of our time has consumed water at the rate of nearly forty gallons per head per day and, to first appearances, the whole of this vast supply is still available. Such a supposition overlooks the vitally important fact that the greater part of the Metropolitan Water Board supply is drawn from the Thames and not from the underground Chalk. By contrast, the marginal regions of the London Basin draw directly from the Chalk, as, for example, in the cases of the Colne Valley, and Barnet and District Companies. Such undertakings are sensitive in respect both of supply and contamination, to conditions 'up-dip', that is, on the marginal Chalk dip-slopes of the central Tertiary tract. With diminishing central population, the Metropolitan Water Board might presumably have some 'exportable surplus' of water, but it is doubtful whether, either from the technical or from the economic standpoint, such surplus would be disposable at widely separated points more than twenty miles from central London. If direct local supplies for the satellites are to be envisaged, acute problems with complex and distant repercussions may well arise. All the northern satellites would trench upon the supplies of 'down-dip' areas, and in the case of the Hertfordshire 'constellation' of towns (p. 245), might affect

---

[1] After these paragraphs were written the Government decision to proceed with a satellite at Stevenage was announced. To some extent at least the resulting outcry reflects what we must regard as the usual partisan and self-regarding motives. The decision may, however, be criticised on other grounds. Alone of the trunk lines north of London the [former] L.N.E.R. main line is bottle-necked by a stretch of double (as distinct from four-line) track through the tunnel at Hadley Wood and Potters Bar. This places a severe limitation on local rail services to points beyond. The choice of Stevenage as a first satellite carries with it, as necessary corollary, the improvement of this transport link, or alternatively, what might be easier, the provision of through services on the loop line via Watton-at-Stone and Hertford.

the sources of the New River, still contributing water to central London. No such acute problems would arise in the cases of White Waltham and Meopham, but the Wealden satellites, Crowhurst and Holmwood, would make heavy inroads on the reserves of an area not naturally well endowed with water.

All such water-supply problems are, in detail, technical, and will require the detailed scrutiny of H.M. Geological Survey and of the undertakings directly involved. It seems, however, a fit matter for comment that the Plan is silent on this topic ; its map showing water-supply areas suggests a major problem which is neither posed nor answered in the text.

3   *The Physical Basis offered by the Plan and its Defects*

So far our commentary has concerned itself with the larger facts of structure which condition the environment. The size of the London region, properly so called, the general features of its hydrology and the possibilities of discovery of coal or oil in the area, all reflect the broad elements of structure, the basic anatomy of the region, and its regional frame. But our scrutiny of the environment must go further than this. It is a patent if not uncommon error to regard the London Basin too simply, as an axial tract of rather featureless country on the .Eocene beds, margined by simple and undifferentiated Chalk uplands. Such is the picture of the small-scale atlas map or the textbook diagram. It conceals behind its over-simplified lines a wealth of variety in the form and constitution of the country. This fact is emphasised by Professor Stamp in his chapter on Land Classification. He there distinguishes in broad terms the areas of river alluvium, gravel and loam terraces, the clay lands, the sandy or gravelly heath-covered lands, the chalk lands, the diversified terrain of the Lower London Tertiaries in the southeast of the Basin, and the tracts of boulder-clay and outwash gravel north of London. His account is illustrated by an excerpt from the published land classification map of the country as a whole, and its chief features may be readily summarised :

1   It distinguishes large areas of category 1A (first-class land, dominantly arable in 1939), comprising the Thames-side gravel and loam tracts and the sandy to loamy Lower Tertiary belt of Kent.,

2 Category 2A (good general purpose farm land) is chiefly in evidence on the Essex boulder-clay.

3 Category 3G (the first-class grass lands) is nowhere present in large continuous stretches, but figures with poorer grass on the estuaries alluvium of the Lower Thames.

4 Category 4G (good but heavy land, subject to restricted working periods and latterly chiefly in grass) is shown as occupying some of the clay country north of London and elsewhere.

5 The clay country of Essex, south of the boulder-clay edge, is shown as 6AG, medium quality farm land, with mixed crops and grass. This same category is applied to a large part of the physically very different 'Bagshot country', apart from its summit areas of 'poor quality light land' (9H).

6 The chalk lands are classed on the whole of mixed type, those of the North Downs being rated rather higher than most of the Chiltern country, where lowland heath and related types (9H) are widely present.

No-one familiar with the London countryside will question that the classification thus applied brings out some salient characteristics of the metropolitan environment. In its nation-wide aspects, indeed, the scheme of classification appears to be sound and useful and inevitably preferable at the moment to the, as yet, very incomplete classifications of pure pedology applied to cultivated soils. It by no means follows, however, that it is entirely adequate as applied to a relatively small and vitally important region such as the London area. It may be useful, therefore, to point out some seeming anomalies as they appear to one closely familiar with the country.

In the first place it must be noted that there are local discrepancies between the published National map and the excerpt printed in the Plan. Thus, much of the Vale of St Albans, and part of adjacent boulder-clay country are apparently grouped as 1A on the excerpt and 2AG on the full map. In fact, grounds are not lacking for grouping both these areas as 1A, though the soils depart somewhat widely from the type represented by the Thames valley and north Kent loams. But no indication is at present before us as to whether the change is designed or inadvertent, and the matter should evidently be clarified.

Passing to more general criticisms we remark that the map on occasion makes strange bed-fellows of very contrasted regions, or alternatively, fails to bring out basic similarities. Thus, the Epping Forest area is grouped with the very different Beaconsfield country in a 'mixed' category, while the 4G land of the heavy clay country of south Hertfordshire is extended, with dubious warrant, into the 'drift country' west of Hertford. Further, as noted above, the physical continuation of the south Hertfordshire country in Essex is placed in a different land classification category, though relief, geology, soil, and settlement, the whole complex of the physical and cultural landscapes, are closely similar in the two cases. Another arbitrary and indefensible line is the western boundary of IA (or 2AG ?) land against the Chilterns, west of St Albans. Near, and to the east of, the latter place, the line has very properly followed the boundary of the gravel and chalk plateaux, but westwards it deviates irrationally from this fundamental boundary and swings down towards Watford, leaving large tracts of gravel country within the chalk terrain of the Chilterns.

It would be tedious and unprofitable to enumerate such discrepancies at length, and ungracious not to concede to the authors of the map a right to their opinion on what must inevitably be a balance of evidence. Nevertheless, in an area so important as the environs of London, it is desirable that alternative interpretations should be given due weight, and vital that the *basis* of the Land Classification map should be closely scrutinised.

The nature of the basis is not far to seek, and from it flow most of the errors and discrepancies of which we are complaining. It is the land-use pattern of the decade before the Second World War (1930–9) which has controlled largely, if not solely, the drawing of the boundaries. Through the courtesy of Dr E. C. Willatts, the writer has been enabled to examine the original land utilisation maps, with land classification boundaries inserted. With the explicit reservation that these boundaries delimit land-use regions, or at least 'cultivation-pattern' regions, and not land-classification regions, complete assent can be given to most, but not all, of the lines drawn. But the reservation is all-important ; land-use regions are manifestly not necessarily equivalent to land-classification regions. They reflect a distribution which, if relatively stable, is none the less, in a longer view, temporary. No-one can suppose that the lines would have been similarly drawn fifty or a hundred years ago, nor can we presume that they are immutably fixed for the future. It is instructive, in this regard, to compare Caird's well-known map of 1850,[1] with the land-utilisation map of our own day. The latter shows the dominantly arable area of eastern England as terminating southwards in east Hertfordshire and north Essex. Caird included the whole of Hertfordshire in the main arable area, as well as Middlesex, the whole of the Weald, Berkshire, Hampshire, south Wiltshire, and Dorset. The subsequent general decline in arable acreage is a familiar and oft-discussed phenomenon. The bodily retreat of the main arable edge in southern England has received less attention. At the earlier date, London lay *within* the arable domain. In our own day it lies beyond it, though near its margin. The London countryside and the South Country in general are thus marginal in respect of dominant arable cultivation ; they have been and will continue to be sensitive to changes in economic, and, we may probably add, in climatic conditions. Point is given to this consideration by the war-time utilisation of the clay country of south Hertfordshire. Given over almost entirely to poor pasture in 1939, the war years brought a notable revival of arable cultivation. Excellent pedigree wheats have been grown here, especially on southward-facing slopes, though at the cost of considerable liming and improvement in field drainage. We must note, in fairness, that the Land Classification map tacitly recognises this fact in its general characterisation of category 4G, and it is widely recognised that the possibility of prompt and speedy mechanical ploughing has generally increased the arable potentialities of such land. It might indeed appear, nevertheless, that the war-time extension of wheat growing here was essentially 'uneconomic', and that a relapse to pasture could not long be delayed in any likely conditions of the world market. At the moment of writing, however [February 1946], this can hardly be said to be the prospect. In any case, the general point at issue must be emphasised ; if we are classifying land and not past or current utilisation, we should surely deal with potential use, and this involves study of the intrinsic qualities of the land, of 'site', water-relations and relief as well as surface geology. The land-use pattern of any age reflects economic environment as well

---

[1] Reproduced in Clapham's *An Economic History of Modern Britain* (C.U.P., 1930), p. 467

as physical land character. The problem of the individual farmer, or the designer of a national plan for agriculture is a literally economic problem, given the physical and technical background, 'the disposal of scarce means between alternative uses'. Land classification must rest on bases less unstable and liable to change.

The question at issue is, indeed, part of a broader problem very relevant in the whole general context of planning—the difference between 'mapping' and 'making a map'. It is the difference between definitive lines which different workers will place similarly, and secondary or subjective lines which represent an interpretation. The boundaries of a physical, geological, or 'land-utilisation map' are definitive within narrow limits, but the generalisation of land utilisation in a specific historic phase does not yield such lines even for its own time, still less for any other. It is a secondary document not to be preferred to primary sources, however suggestive and valuable it may be.

It is for these reasons that one must regret that a land-classification map was preferred to physical, geological, and other maps in an attempt to present a picture of the regional environment of London. It may succeed in relation to the specifically agricultural purpose, but it fails in the wider context, and if it appears to be an economy in presentation, it is an economy purchased at heavy cost in accuracy and objectivity.[1]

4 A Suggested Physical Basis for the Plan and its Relation to the Proposed Zonation

The foregoing strictures evidently impose upon us the duty of suggesting some other basis of regional classification which may

[1] It may be added that these considerations apply with even greater force to the recently published 'Types of Farming' map. That areas may be broadly classified on a basis of the dominant types of farming enterprise is not to be gainsaid. What is highly regrettable is that such areas should be delimited by clean-cut and over-simplified lines. One states with assurance that, both in the London district and in other parts of the country, many of the LINES appearing on the map have no geographical significance whatsoever, and the grouping together of dissimilar terrains produces results which border on the bizarre. Nor is it any defence to say that the map is not concerned with terrain as such ; it does violence to the proved relationship between terrain and farming practice. It is not a map indeed, but a crude and ineffective cartogram, based on criteria so loose and vague as to render the results either self-evident or ludicrous.

better serve as a background to the London Plan. In 1932 the writer published [1] a scheme of sub-regions for the London Basin, which is reproduced in slightly amended form in Fig. 30. The pattern presented is significantly similar to that of the map of land-classification, and it will be well to make clear wherein the difference lies. It resides in the fact that land-use pattern was not explicitly considered in drawing the regional boundaries. These boundaries reflect the facts of geomorphology and surface geology, considered in their general environmental context— i.e. as determinants of human settlement and occupancy. In so far as these regions show relationship to recent land-use, the fact is important and duly to be noted, and in many, though not all, instances, it lends support to the land classification categories adopted by other workers. But it must be emphasised that these regions would remain legible quite independently of any change in land-use modes. If Macaulay's hypothetical New Zealander, in some unimagined future, after musing on the ruins of London Bridge, were to make a pensive traverse either northwards or southwards from the Thames, he would, given an 'eye for country', discover these same regions whatever the prevailing land-use mode, or the degree of relapse towards wild vegetation. The regions are, or rather were, in some sense *pays* in the non-metro-politan past, and as such are significant to the historical geographer. The historian or archaeologist of the distant future reconstructing the lineaments of our own culture and civilisation would find them not less significant. They are, indeed, the fundamental units out of which both the physique and personality of the larger region are constituted.

We cannot here present a full and formal description of the sub-regions, but a brief commentary may serve to make their nature and relations clear.[2]

The Sub-regions of the London Basin. The Chiltern plateau is really the largest of the regions, and by no means the least distinctive. It is a Chalk upland in a state of youthful or sub-mature dissection, scored by sub-parallel valleys, largely dry ;

[1] *Geography* **17** (1932), 99 ; and Essay No. 10, p. 150 of the present volume
[2] Although the planning area extends into the Weald, no account is offered here of the Wealden sub-regions involved. They can hardly be discussed in detail except in relation to the Weald as a whole, and their broad features are obvious and well known.

these converge locally to valley-nodes commonly marked by town sites. The broad interfluvial plateaux are uniformly mantled with clay-with-flints and plateau brickearth, while relict Eocene masses rise, with relatively steep slopes above the general surface. The region retains a heavy woodland cover, especially in its

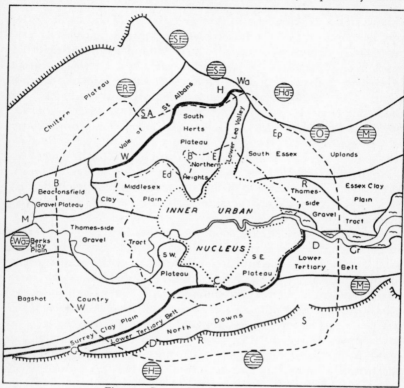

Fig. 30   Greater London Regional Plan

western part. The settlement pattern is not conspicuously nucleated, save in the valley trenches. On the plateaux are what were originally woodland hamlets and large isolated farms, the latter dating in many cases from the time of the Napoleonic wars, which saw a drive for home-grown food reminiscent of our own time. The relief of the Chilterns diminishes both in amount and sharpness of expression from southwest to northeast, but the plateau terminates abruptly in the ridge between Offley and

Knebworth, at the foot of which lies the 'double' Hitchin Gap (Langley and Stevenage valleys). Beyond the latter the undulating boulder-clay plateau begins ; it differs radically in hydrology, soil, and texture of dissection from the Chiltern country.[1]

Along the southeastern margin of the Chiltern plateau we approach the line of the pre-glacial or early glacial Thames valley. Between Henley and Bourne End the Thames still follows this line, but east of the latter place the former activities of the river, and of the tongue of ice which entered its valley from the northeast, are marked by a complex drift belt. In topographic expression and general soil character this falls into two distinct parts. In the west, the Beaconsfield plateau is thickly and uniformly gravel covered, and for the greater part underlain by Eocene sands and clays. The gravel sheet comes on abruptly at 350–400 feet, and the general surface falls southwards in well-marked terrace-like stages to a gravel bench at about 200 feet O.D., immediately overlooking the Thames valley at Slough and Windsor. The country is heavily wooded and sparsely settled. It is not to be regarded as similar to, or part of, the Chiltern plateau (vide Par. 223).

The Vale of St Albans, continuing the drift belt, is a broad trough lying between the margin of the Eocene outcrop (the so-called Tertiary escarpment) and the edge of the gravel terrain on the Chiltern dip-slope, maintaining a curiously straight course and a uniform elevation of about 400 feet. The trough is drained westwards by the Colne and eastwards by the Lea, but is manifestly a pre-existing depression which these streams have come to occupy. Large parts of the floor, especially south and east of St Albans, are remarkably flat, marking, in all probability, the bed of a temporary pro-glacial lake. It is this feature which favoured aerodrome construction as at Radlett and Hatfield. The soil conditions are much more favourable than those of the Beaconsfield plateau, since tracts of boulder-clay occur (marking ice-protrusion from the northeast) as well as large stretches of loam or brickearth, not separately distinguished on the geological maps. Moreover, the gravels, lying here at a lower level, and within reach of the water-table, yield medium to first-class arable

---

[1] *cf.* S. W. Wooldridge and D. J. Smetham, *Geog. Journ.* 78 (1931), 243 ; and Essay No. 11, p. 171 of the present volume

soils. By virtue of these facts the region marks the first on-coming of high-grade arable country north of the Thames valley. Within it, five considerable towns are located, Hertford and Ware at the eastern end, Watford and Rickmansworth at the western end, while St Albans is centrally situated on the northern border. Welwyn Garden City and Hatfield, latterly much expanded, lie on its margins.

The northern edge of the south Hertfordshire plateau, which overlooks the Vale of St Albans between Watford and Ware, is a clearly marked regional boundary. Beyond it lies heavy clay country, but a well-developed gravel-covered summit-plane dominates the district at an elevation of about 400 feet. In the west, there is little more than a narrow discontinuous ridge separating the Vale from the clay plain of Middlesex. Near Barnet, however, the plateau broadens, and here we find island 'gravel tops' as at Ridge and Shenley, separated by wide and open clay vales. Similar conditions prevail along and east of the Great North Road, though here local cappings of boulder-clay introduce a new soil element. The eastern part of the area presents conditions of surface, soil, and vegetation cover almost identical with those of the higher parts of Epping Forest.[1] The settlements throughout the area (excluding latter day accretions such as Boreham Wood and 'suburban' Potters Bar) are on the hill or plateau tops.

The geological and morphological characters of the south Hertfordshire plateau extend southwards of High Barnet in the prong of hilly country which includes the high ground at Totteridge, Mill Hill, Finchley, and Muswell Hill, and the outliers of Bagshot sands at Hampstead and Highgate. We have distinguished this tract as the 'Northern Heights'; it has been an important feature in conditioning the growth of London and remains a clearly marked factor in its 'social geography'.

Beyond the broad trench of the lower Lea valley the 'site qualities' of the south Hertfordshire plateau are continued in only slightly variant form in the south Essex uplands. Lying south of the boulder-clay edge at Epping, Ongar, and Chelmsford, this

---

[1] It is here, under the extended name Broxbourne Woods, that the Plan wisely proposes to conserve a tract of almost unspoiled country within the Green Belt Ring.

tract includes the Epping Forest country and, farther east, a series of broad clay vales with widely separated island 'tops', largely outliers of Bagshot sand, as at Havering, Brentwood, etc. Tongues of ice from the north protruded between the hills of this belt, leaving witness in sheets of boulder-clay, now somewhat dissected, on the middle slopes.

Of the original characters of the Middlesex Clay plain we need say little, since most of it now falls within the London boundary. It is a featureless, gently undulating tract ranging in general elevation from 100–200 feet, but studded with conspicuous residuals such as the Harrow-Sudbury mass and the minor eminences of Kingsbury, Horsenden Hill, etc. These give what little character there is to the physical background of northwest London. The Essex Clay plain, east of Romford, lies nearer the Thames and is flatter in general aspect and with fewer residuals, though the Laindon Hills are a close analogue of Harrow Hill.

The Thames, in its post-glacial reincarnation, resumed its earlier tendency to shift steadily, or by stages southwards, and accordingly the major spreads of terrace gravels lie on the north bank. This Thames-side gravel tract forms a wide plain west of London, descending from about 100 feet to the river's edge. Eastwards, beyond the Brent, it constitutes a broad gravel ledge (essentially the Taplow or middle terrace of the Thames) at 40–80 feet above the river. This continues through central London, giving a high north bank to the river ; only on the eastern outskirts does the main ledge recede from the river, leaving room for a lower gravel surface, the flood plain terrace, which in turn gives place to the river-side alluvium.

While the Thames has moved on the whole southwards, the lower Lea has moved eastwards, under structural control, abandoning a tier of terraces on its western side, of which the chief is again the middle or Taplow terrace. The valley as a whole remains a dominant feature in the geography both of London itself and of its countryside. The proposals of the Plan for the rescue and integration of this major feature are bold, but wholly commendable.

It is needless to deal here at length with the important rôle of the Thames and Lea gravel tracts in directing the earlier stages of London's layout, in providing supplies of water at shallow depth, and, with or without a brickearth covering, affording, perhaps, the highest quality and most easily worked arable land in the London Basin.

London's southern countryside presents, on the whole, a simpler physical picture than the northern area, but one not altogether easy to resolve into valid regions. The main Eocene edge is an unmistakable line throughout. West of Croydon distinctiveness is given by the relatively high regional dip ; the Chalk dip-slope rises sharply from beneath the low-lying Tertiary margin and gives the North Downs a wall-like aspect as approached from the north. The sudden oncoming of the Blackheath beds, in the Lower London Tertiaries, east of Croydon, results in a true Tertiary escarpment traceable thence to Erith. Within the Tertiary tract the positive elements in the relief pattern are, respectively, the Bagshot country and the areas we have termed the southwestern and southeastern plateaux respectively. The first is a dissected sand plateau with gravel-capped tops. The southwestern plateau, between Kingston, Richmond, and Wimbledon, owes its existence to an outlying tract of glacial gravels. The southeastern, or Blackheath, plateau is essentially the upper surface of the pebbly Blackheath beds, dissected along lines determined by minor folding, and covered in part by relict masses of London Clay. Only locally, as at Shooter's Hill and in the Norwood Hills, do these attain or approach the high '400-foot' plateau level which dominates north London ; the general plateau elevation is about 200–250 feet.

The low-lying elements in the pattern comprise the narrow clay plains of Surrey and Berkshire which wrap around the Bagshot heaths, and the main lowland of south London, which, until it merges in the flood plain of the Thames, is essentially the Wandle valley.[1]

[1] It is, of course, possible, though in this case unprofitable, to make a more detailed division of south London than this. The Wandle valley from Croydon via Mitcham and Tooting to Wandsworth, is a gravel tract in some ways comparable with the lower Lea valley, though less dominant in its relief. The Norwood-Forest Hill ridge and its spurs form the core of a London Clay tract extending from Streatham in the west to the Ravensbourne valley ; in some sense this is a 'Southern Heights', balancing the Northern Heights, with a similar rôle in 'social geography'. But in relation to the general urban structure of London there is some warrant for including the Norwood Hills, as implied above, as part of the southeastern plateau.

Beyond the Tertiary escarpment between Farningham and Erith the lower members of the Lower London Tertiaries (principally the Thanet sand) range on discontinuously eastwards, interspersed with areas of Chalk modified by the Eocene relics. This is the western end of the Lower Tertiary belt of Kent— the north Kent fruit zone. It is continued in some sense in the country beyond the main Tertiary edge west of Croydon, where, though the surface is mapped as Chalk and the Thanet sands are thin, there is sufficient residual Tertiary 'skin' to modify the soil favourably for arable farming.

The generic characters of the North Downs plateau are obviously akin to those of the Chilterns subject to the variations imposed by a higher general dip and a somewhat different extent and character of post-Cretaceous cover.

The relation of the Planning Zones to the Sub-regions. On the basic pattern of sub-regions shown in the figure, we have superimposed the fundamental zonal boundaries of the Plan, though with certain modifications. The closely built-up areas of the County of London have been merged with the Inner Urban Ring to form what is termed the Inner Urban Nucleus. The area so defined excludes the southeastern part of the County of London which is of suburban character and building density. Though the official Plans must necessarily recognise administrative boundaries, they may well be neglected in seeking a general picture of the whole urban unit, especially since they are demonstrably outmoded, if not sheerly grotesque. The outer limits of the Suburban and Green Belt Rings are generalised, but not otherwise altered.

Though the map shown is, in some respects, crude and simplified, it makes a concise statement of relationships and poses questions even if it does not, in all cases, answer them. It brings to light the control of urban growth by relief. Inner London was constrained by high ground ; it thrusts northwestward between the southwest plateau and the Northern Heights and, held back by the latter, advances again up the lower Lea valley. Similarly, the main mass of south London as far south as Croydon lies below and between the two southern plateaux. We are, of course, to read into this no compulsion exerted by relief. With local exceptions, the slopes are not steep enough to deter building.

We see, rather, the effect of relief upon surface rail-routes and the tendency for plateau sites with scenic amenity to be retained in private ownership as open spaces. Of these two factors, the latter is more important. The 'mountain-railways' of suburban London climbed in due course to terminal stations at elevations of over 300 feet, as at Muswell Hill and Upper Norwood. But the hill-top site, with its old village and its fringe of well-wooded parkland, now variously employed, is characteristic of both north and south London.

On the outline of suburban London, relief as such has necessarily had less effect, though isolated hills still pleasantly punctuate the pattern. In two instances, however, the suburban boundary coincides closely with a physical line, between Stanmore and Mill Hill, and along the edge of the Blackheath plateau overlooking the Cray valley. In the first instance, the red suburban tide has rolled across the Middlesex plain to the foot of the Hertfordshire hills, pressing through them in the gaps at Hatch End and Northwood, but holding back in notable degree farther east, so as to leave the Totteridge-Mill Hill country green. This last, indeed, is a remarkable piece of chance conservation, bringing wide stretches of undulating pastureland, with timbered hedgerows and war-time wheat fields surrounding two charming and unspoiled villages, within eight miles of Charing Cross. We can trace this chance preservation to the latter-day practice of electrifying the surface lines instead of forging new 'tube radii' ; the country in question lies between the Edgware and High Barnet branches of the old Great Northern suburban system, one electrified and the other shortly to be completed. It is instructive to compare the case of the Piccadilly Underground extension to Cockfosters, which carried the spread of housing in a graceless and unfinished fashion across the slopes and plateaux between Barnet and Enfield, very similar once to the Totteridge country.

The foregoing comments on points of detail may have certain interest for the student of urban geography, but they have little bearing on the future of London. The selective conservation of hill and plateau tops has been a factor in the piecemeal 'planning' of London, but there is no presumption whatever that a similar factor would continue to act unless deliberate general conservation measures, such as the Plan proposes, were taken. Assuming, however, that the Plan is accepted, in essence, our map offers a purview of what it is we are proposing to conserve, not merely in terms of land classification, but in natural scenery. Let us examine the regional contents of the Green Belt Ring by tracing its approximate median line round the Metropolis. On the south, it lies near the crest of the North Downs, and the whole scarp-edge between the Mole and Darent gaps, together with the Vale of Holmesdale and contiguous parts of the Weald, lie within the Ring. This calls attention to the fact that Greater London lies asymmetrically in the London Basin extending beyond its southern, but nowhere approaching its northern, edge. Eastwards, our line traverses the Lower Tertiary belt along the Darent valley, still largely open country, south of Dartford. Thence northward across the Thames to Romford there is little peace or scenic amenity within the Green Belt Ring, though considerable tracts of valuable agricultural land survive around the Ockendons. From Romford, northwestward across the Roding valley to Theydon Bois, Epping Forest, and the Lea valley, lies one of the most valuable sections of the Green Belt Ring ; here the filthy impress of London has as yet left little mark, and one can walk for a dozen miles or more from Noak Hill or Havering to Nazeing, through a pleasant and diversified countryside, and one which might well have been rated higher in the agricultural scale than the land classification scheme suggests.

West of the Lea valley, the Green Belt Ring includes the whole of the south Hertfordshire plateau, the western part of the Vale of St Albans and a part of the lower Chiltern dip-slope. The most valuable stretch of unspoiled country lies between the Lea valley and the Great North Road. Beyond the latter, the network of routeways, present and proposed, and dormitory suburbs as at Potters Bar and Radlett, will go far to spoil the Tertiary country, and the western end of the Vale of St Albans is all too heavily spotted with building ; though retaining high-grade ploughland in considerable amounts. Vital to the preservation of amenity and 'visual solace' is the preservation of both edges of the plateau country. From the northern side, as at Ridgehill and Shenley, the eye ranges to the Dunstable Downs across a foreground of almost 'champion' ploughlands in the Vale and a background of Chiltern 'bocage'. On the southern side on the ridge between Barnet Gate and Harrow Weald Common, there is, as we have seen, a well-marked limit to London. The ground rises from the suburban edge in bold slopes to bare or tree-crowned summits, and the southward view extends to the North Downs, broken by the cleft of the Mole Gap. Beyond the Vale, the Green Belt Ring boundary includes a choice and characteristic fragment of true Chiltern country—'Metroland' to the Londoner—around Chipperfield and Chenies.

Our 'median' traverse would carry on thence southward, west of the Colne valley, across the Beaconsfield gravel country, still rich in scenic attraction, and across the lower gravel country to the Thames at Staines. Here, and again southward to Leatherhead, stretches of true country are less extensive ; we perforce traverse the 'ribbons' of the Great Western main line at Slough, of the Southern Railway at Woking, and of the settlements of the Tertiary edge between Epsom and Effingham. Only west of the latter is it possible to see in their natural relationship the Bagshot sand country, the narrow vale of the London Clay and of the Chalk upland beyond it.

All this at least, no inconsiderable series of assets, the Plan would preserve for us within the Green Belt Ring, and though blemishes of former 'non-planning', too numerous to mention and impossible to remove, are all too frequent, London thus safeguarded would retain a countryside, or a series of countrysides, as its immediate frame and setting. Leaving aside the debatable questions of the satellite towns and of the ultimate fate of the Outer Country Ring and of the London region (*sensu lato*), it is difficult to resist the conclusion that the Plan involves minimum demands in respect of the Green Belt Ring and makes something like the best of a thoroughly bad legacy of insensate former growth. No valid case whatsoever can be made against such minimum demands. In this regard, however, our map may serve one more purpose. Let us suppose that self-regarding interests or the compelling pressure of temporary housing difficulties were to bring about some modifications of the Plan. It is easy to see how, on apparent balance of considerations, reinforced by the customary so-called 'economic' arguments, it might be sought to justify local trenchings upon the Green Belt Ring. The figure directs attention to some of the more 'sensitive' points or areas which might figure in such debate. The general principle which emerges is as follows ; the northern and southern sections of the Green Belt Ring perimeter run on the whole with or parallel to the regional boundaries ; they should be strictly adhered to. On the east and the west the Green Belt Ring runs transversely to the natural elements of the terrain, and modification of its boundaries is likely to make less difference to the regional 'make-up' of the country preserved. We must note, however, that if scenic amenity is here less in question, the case of agricultural use is perhaps even more pressing, since what is involved is further loss of the Thames valley loam and gravel lands. Reverting, however, to the northern and southern boundaries, it is clear that in either area, and especially in the north, the extension of surface electric lines with uncontrolled building, would rapidly consume some of the most distinctive parts of the London countryside, the clay country of Hertfordshire and the chalk country of Surrey.

5 *Conclusion*

We have attempted to indicate, within necessarily narrow limits of space, the lines on which a regional description of the London countryside might be written. It is to be hoped that British geographers will find means and opportunity to write much more adequate accounts of their parts of the country if a widespread policy of planning is now to be carried out. It is still lamentably true that the detailed regional geography of this country remains to be written. In conclusion, however, we may perhaps be permitted to point out that the bases of full description in the London Region are not yet to hand, in that the primary six-inch geological survey of the area is still incomplete, while soil survey in the proper sense is hardly yet begun. It may well seem, in the future, a curious commentary on the mind of our generation that we have been forced by pressure of facts to confront the task of town and country planning while still lacking in many categories the basic surveyed data on which such planning should be based.

15

The Working of Sand and Gravel in Britain

A Problem in Land Use

THE inroads made by sand and gravel working upon the land surface of Britain attained formidable dimensions between the Wars, and war-time needs accentuated the difficulties of a situation which threatened to become unmanageable. Sand and gravel, reckoned by bulk, have for some time been second only to coal among the minerals worked in this country. Before the war of 1914–18, production was under two million cubic yards a year, and the extraction of gravel could hardly be termed an industry, for pits were small and scattered and were worked by hand. Output rose steeply after 1919; by 1938 it had reached a figure of twenty-five million cubic yards, and our needs for future years are estimated by the Ministry of Works at thirty million cubic yards.[1] Converted to acreage on the basis of an average yield of 20,000 cubic yards to the acre this would mean an annual consumption of 1500 acres of land. Making every possible allowance for after-treatment of sites, it involves at least temporary, and probably permanent, damage both to amenity and agricultural production. Yet, clearly, the present need is not to be gainsaid. The uses of sand and gravel are many, but the increased output is essentially due to the increasing use of concrete, for which sand and gravel provide the necessary aggregates. If we take account, too, of building-sand and of gravel for road-making it is clear that 'surface development' depends upon the provision of adequate and reasonably cheap supplies.

The problems involved have engaged the attention of the Government, and in 1946 the Minister of Town and Country Planning appointed an Advisory Committee of which the authors of this paper were members. The Committee has published its

---

[1] A cubic yard of gravel weighs about 1¼ tons

General Report [1] and a more detailed examination of the problems
of the London area. It is now at work on similar reports for the
remainder of the country. It is not our purpose to repeat or discuss
the recommendations of the Committee, still less to anticipate
its future findings, but rather to bring out the general shape of the
problem. The rôle of the geographer in this and kindred matters
requires careful limitation. That a number of strictly geographical considerations are relevant is obvious enough. But
there are others. The appropriate measure of 'control' is, for
example, a question of political philosophy and Government
policy and also one of administrative machinery. With none of
these aspects has the geographer, as such, anything to do. His
findings as far as they remain, as they should remain, accurate
and objective, are as relevant for the convinced 'planner' as for
his opponent. But the land patterns of 'surface development',
agriculture, and this form of mineral working are facts of geography, the expression of geological, technological, and economic
factors, and no decisions, not even the basic one to control or not
to control, can be taken without an analysis of the existing
geographic pattern.

The Geological Background

The primary factors governing the problem derive from the
facts of geology. It is worth reflecting that there are very few
parts of the land surface of Britain which might not be drawn
on for mineral products of one sort or another. But an important
distinction separates the production of, say, gypsum or iron ore,
which can only be worked in clearly defined and narrowly
localised areas, from that of widely spread materials like sand,
gravel, and brick clays. Moreover, brick-making, once widely
diffused in small workings, has become increasingly concentrated
and the possibility of such concentration lies in the thickness,
homogeneity, and 'predictability' of the deposits used. By
contrast, the main source of sand and gravel is the irregular
mantle of Pleistocene drift deposits. There are important local
exceptions, for constituents of the 'solid' rocks, like the Bunter
pebble beds, or the Lower Cretaceous and Eocene sands,
provide important supplements, and where such deposits are

[1] *Report of the Advisory Committee on Sand and Gravel* (H.M.S.O., 1948)

worked their generally greater thicknesses provide higher yields per acre than the drift deposits. There are, however, few areas of the English lowland which do not derive about half, and generally more than three-quarters, of their sand and gravel from drift deposits. The latter comprise two significantly different groups : (1) The glacial or plateau gravels, marking glacial outwash beyond the limits of the ice-sheets, or in the south, similar sheets of sand and gravel distributed by distant ancestors of the present rivers beyond the confines of the later valleys ; (2) at lower levels and derived in large measure from these earlier deposits are the terrace or valley gravels, clearly attributable to the present rivers or their immediate ancestors. One may speak of these two classes as the 'high-level' and 'low-level' deposits respectively, but the distinction between them is not primarily altitude but age, origin, and, for our present purpose, character. It is frequently difficult to distinguish between the higher terraces and the lower spreads of plateau or glacial gravel, but it is generally true that the latter are less regularly bedded and less well sorted, and for this reason alone the terrace gravels are in general easier and more profitable to work. Washing is essential in preparing concrete aggregates, and the terrace gravels have necessarily undergone natural washing in repeated deposition and redeposition of the same material.

In the relative bulk and extent of the 'high-level' and 'low-level' deposits a distinction can be made between the northern and southern part of the country. Avoiding the still actively disputed field of detailed Pleistocene chronology, we may distinguish the Older Drift, marking a time (or times) when land ice extended as far south as the Thames valley, from the Newer Drift, the product of much later and more limited advances following the retreat (and probably the complete disappearance) of the earlier ice-sheets. The edge of the Newer Drift trends irregularly across the country from the Yorkshire and Lincolnshire coasts to the Welsh border and South Wales. To the south, and notably in the valleys of the Thames and the Wessex rivers, terrace gravels are well developed and were in fact deposited while the North Country was still covered by ice. In the north there has not been time for the development of a full terrace sequence since the ice of the Newer Drift withdrew. Thus the great Yorkshire rivers and the upper Severn show no terrace sequence comparable to that of the Thames or the Salisbury Avon ; the gravels adjoin the streams and extend only to a limited height above them.

A review of the changes in the location of gravel working during the last thirty years shows the tendency of the industry to move into the valleys. One reason for this we have noted, the superior quality of the valley gravels. Of equal importance is the introduction of 'wet working', described below. The 'flood-plain' gravels of the rivers, the lowest member of the terrace series, are water-logged, with the permanent water-table but a few feet below the surface and the gravel can be pumped or dredged. Similar conditions occur on the higher terraces where the gravel rests on impermeable 'country-rock' such as London Clay or Keuper Marl. The methods employed and the demand for gravel thus place a premium on the riverine lands.

Geological and physiographic facts lie not less certainly behind the two other forms of land-use which compete for these same tracts. We need not here pursue the theme of settlement and urban growth far back into history. The attraction exercised in the earlier stages of settlement by river sites and the adjoining tracts of terrace land is a familar and recurrent theme in the historical geography of this country. A small-scale dot-map of early Anglo-Saxon settlements might be mistaken, at a casual glance, for a map of modern gravel pits ; the seeds of our towns were often sown in the gravel fields. More germane to our problem has been the tendency for the town growth of the industrial age to concentrate upon the gravel terraces. London, north of the Thames, expanded east and west along what has been called the 'London gravel ledge' and did not extend beyond it before the coming of the railways. Later the supposed attractions of a 'gravel sub-soil' and the advantages of level ground for lay-out were sufficient to ensure a wider expansion of relatively low density residential property and light industry on the terraces. The loss of agricultural land involved has often been the subject of dismayed comment.[1] Only more recently has it come to be realised that urban expansion has also sterilised wide areas of

---

[1] See, for example, E. C. Willatts, *The Land of Britain, Part 79 : Middlesex and the London Region* (Murby, 1937), pp. 192 202 ; also *Greater London Plan, 1944* (H.M.S.O., 1945), pp. 85–6

gravel, not less necessary than food-producing land to the continued life of the towns.

Gravel and Agriculture

Few general statements can be made concerning the agricultural quality and value of sand and gravel soils. The parent material varies so widely, particularly in respect of the proportion of finer particles, which affect its water properties, and the circumstances of 'site' are themselves so variable, that the resulting soils include some of the best and some of the worst that the country can show. Nevertheless some generalisation is possible. The distinction between 'high-level' and 'low-level' deposits is as valid for soil quality as it is for gravel quality. Much of the glacial and plateau gravels and some of the higher terraces under a semi-natural plant cover present characteristic stretches of the oak-birch-heath association of the plant ecologists ; the soils are heavily leached in respect of bases and sesquioxides, and attain or approach the character of true podsols. The cultivated portion of such land is often in the sub-marginal category. By contrast the true riverine lands provide some of our best soils. Perhaps the most important, in the south, are the areas in which gravel carries a brickearth cover. This is for the most part undistinguishable from the European loess, or the limon of northern France. Much of it is certainly wind-blown, though some may have been subject to secondary water-deposition. In mechanical constitution an almost perfectly balanced loam, it gives rise to some of the finest arable soils in Britain, essentially identical with *les sols de culture facile* of Vidal de la Blache. Some may count it an ironic commentary on the mind of an over-urbanised people that our name for the deposit should be 'brickearth'. It provided countless millions of the characteristic yellow London 'stock' bricks in the earlier stages of the growth of the Metropolis. Soils derived from brickearth provided the pick of the now largely vanished orchards and market-gardens in west Middlesex, and there has been little dissent from the view that surviving brickearth tracts should be reserved for agriculture, especially since brickearth in any thickness is an objectionable overburden to the gravel operator.

Other areas adjoining the rivers have an alluvial cover over gravel and are highly prized as pasture land. Here again, the alluvial cover, if thick, is an obstacle to gravel working, but, especially in the North Country, many areas mapped as alluvium comprise or contain high-quality gravel.

Where free from alluvial or brickearth cover, gravel areas often rank high as fruit or market-garden land by virtue of long-continued artificial upgrading ; the humus content has been raised by green-manuring, by grazing stock on the land, or by heavy and continuous dressings of organic manure. In this way soils have been produced which show a notable flexibility in utilisation and which make profitable high capitalisation and a heavy application of labour. As a result, in a general scheme of land classification such as that devised by Professor Dudley Stamp, the whole of the gravel field in such an area as west Middlesex can be graded as first-class arable land, whether it is surfaced by brickearth or not.

As far as present circumstances justify the active protection of our resources of home-grown food, these facts may seem to point to a very simple policy of land allocation. From the agricultural point of view it would be highly desirable if gravel-working could be wholly relegated to the high-level deposits. We have already indicated why such a policy is impracticable. To pursue the matter further, it is necessary to examine more closely the present structure and character of the sand and gravel industry.

The Gravel Industry

As we have seen, the industry grew from very small beginnings. Land was at first relatively cheap, plant was comparatively cheap and easy to obtain without delay, and there was a rapidly expanding market. Despite the low value of the product, therefore, some substantial profits were made, particularly in geologically and economically favourable areas. But there were also many failures, due to geological ignorance or inadequate testing of the ground, and in course of time the more successful producers acquired more pits and plant, in the same or in other areas. Limited liability companies replaced one-man businesses and firms increased in size. Further, gravel-producing came to be a subsidiary activity to other businesses. Coal merchants and transport firms, for example, opened gravel pits as a side-line, perhaps at first to ease seasonal under-employment ; public-works contractors entered the business as an obvious way of securing an important raw material.

At the present time just over one-third of the national gravel output is produced by large firms numbering only 4 per cent of the whole ; a further 13 per cent by somewhat smaller firms numbering 5 per cent of the whole, and the remaining 51 per cent by small firms numbering 91 per cent of the whole.[1]

Our geological introduction has shown that excavations for sand and gravel may occur in a variety of geological and physical situations. As one of us has pointed out in a previous paper,[2] the physical form of any quarry depends broadly on (1) the thickness of the bed being extracted, (2) the thickness of the overburden, (3) the nature of the rocks comprising the over-burden and the mineral bed, and (4) the relation of the excavation to the water-table ; a fifth factor, the nature of the apparatus employed, is also of considerable significance.

A large number of gravel pits fall within the first and simplest class of quarry, that in which a relatively thin bed is removed from beneath little overburden, with the excavation quite dry. Such pits are commonly worked with face shovels or dragline excavators, and the overburden may be dumped into the worked-out portions of the pit.

The second class of quarry, in which a thin seam of mineral is extracted from beneath a thick overburden (as in the Northamptonshire ironstone field and in opencast coal excavations) has no representatives in the gravel industry, for the obvious reason that the cost of removing the overburden would raise the price of the gravel far beyond its present economic value.

The third class, in which a thick bed is removed either by excavating vertically downwards or by advancing horizontally into the face of a hillside, has relatively few representatives for gravel beds are seldom of great thicknesses (over 30 feet) and seldom form escarpments. Glacial sand and gravel deposits of morainic or esker-like form may occasionally reach considerable

---

[1] *Report*, p. 27
[2] S. H. Beaver, 'Minerals and Planning', *Geog. Journ.* 104 (1944), 169

thicknesses ; a pit near Wrexham, in thick glacial accumulations at the junction of the Cheshire plain and the Welsh massif, once had a face 120 feet deep,[1] and over 80 feet are still worked in the vicinity. Thicknesses of over 40 feet are worked at several pits in glacial gravels in the north of England, especially in north Yorkshire and County Durham. The Bunter pebble beds too, in the West Midlands, often yield considerable thicknesses of sand and gravel, and the pits, like those in the thick glacial gravels, are commonly cut into the face of the hillside rather than dug directly downwards. The finest Bunter gravel pits are in Staffordshire, where, as near Cheadle, a face of about 150 feet is known to have at least another 70 feet of workable material beneath it ; similar pits are to be seen in and near Cannock Chase. These deep-faced gravel pits are commonly worked by face shovels, sometimes in more than one level ; their output (rising to 100,000 cubic yards per annum or more) and the size of the plant may be considerable. They obviously present, like quarries of similar type in other rocks, a difficult if not insoluble problem of 'restoration'.

The fourth class of quarry, in which the excavation penetrates below the water-table and the mineral is extracted partly or entirely from beneath the water, is virtually confined to the sand and gravel industry, though many quarries in other rocks which go below the water-table are kept dry by pumping. It is these 'wet' pits which are characteristic of most of the lower-lying terrace gravels of the Midlands and south and of the alluvial and fluvio-glacial gravels of the north. The gravel in 'wet' pits may be excavated by a dragline or bucket-chain stationed on the bank of the lake and dumping its load on to a belt conveyor, or into Jubilee trucks or dumpers or even into lorries, for transfer to the screening plant ; or it may be won from a floating pontoon which carries either a bucket-dredger or a suction pump discharging into a barge. Clearly this latter method can only be used if there is a sufficient depth of water (and so of gravel) to float the pontoon and barges. If the gravel be too deep or composed of heavy materials with many large stones, pumping may be difficult or impossible. In the predominantly flint gravels of the Thames

---

[1] C. B. Wedd, B. Smith, L. J. Wills, 'Geology of the country around Wrexham', Part 2, pp. 142-3, *Mem. Geol. Surv.* (1928)

terraces which attain a thickness of 20 feet or so and contain very few large stones, the use of suction pumps is very common. In the north of England, however, for example in west Yorkshire, where thick fluvio-glacial gravels are worked and the predominant pebbles are of Carboniferous limestone and sandstone, or igneous rock, many stones and boulders are of such a size that they would obstruct a normal suction pump. Here the Sauermann Conveyor (or 'tower dragline') is frequently employed. Because of the capital expenditure required for plant, 'wet' pits usually have a substantial output, and the lakes created may cover scores of acres.

Technical and Economic Factors

Technical development in the last three decades has enabled the gravel industry to cope efficiently with the variable nature of the deposits. But, to an ever-increasing extent, engineers are demanding higher-grade aggregates, corresponding more closely to specification as to size, proportion of particles of different sizes and so on, and the result has been to increase the complexity of the washing, crushing, and screening plant and to encourage the production of a great variety of sand and gravel qualities. This in turn necessitates larger capital expenditure and larger outputs to cover overhead costs. The gravel industry has indeed advanced a great way from the production of 'hoggin' by hand, and is now probably the most highly mechanised extractive industry of all. Another important result, as yet little appreciated outside the industry, is that many deposits which might have been worth working twenty years ago, or even later, are not now economic (though they may have been used in the hectic rush of airfield construction during the Second World War), since by their geological nature they are unsuitable for the production of high-quality aggregates.

To the casual observer, a gravel pit is apt to wear an ugly and unfinished aspect, more so than most other types of quarry. What is easily overlooked is its ingenious adaptation to the shape and size of the working site and to the varying character of the deposit. Each pit presents its own problems and with the elaboration of the process the land requirements of a production unit have increased. The heavy capital expenditure on plant requires a long 'stand' on any one site. It cannot be assumed that the same plant will avail for another, and a period of not less than fifteen years' working is now normally regarded as the economic desideratum. Space must be found not only for plant and storage, but for the disposal of fine washings and, if after-treatment is in question, for the stacking of filling or surfacing material. In general, sites of at least fifty acres are required though under suitable conditions areas half this size may be efficiently worked. The sub-division of sites by roads, railways, rivers, or even public footpaths presents acute problems.

Some of the most important limitations on the choice of site result from the inherent variability of sand and gravel deposits. Many local 'failures' in quality or thickness derive from the original conditions of deposition, others from subsequent disturbance of the deposits. Terrace deposits generally become thinner towards their rearward edge ; on the forward or river-ward edge, the bluff bounding the lower succeeding terrace may expose the underlying country rock, an obvious line of failure. But the gravel may be continuous at the surface because of downwash or slipping, and in some cases secondary accumulations of gravel below the bluff give rise to locally increased thicknesses near such areas of thinning. Another common and troublesome phenomenon is the later channelling of a deposit and the filling of the channels with finer-grained material which is worthless overburden to the operator. In the Bunter pebble beds the gravels are generally lenticular and often with too great a thickness of sandy material for economic working. The industry has availed itself very little of geological advice and many operators are unaware of the most elementary principles of geology. Some past working has proceeded on a wild-catting principle and a great deal of avoidable damage to amenity has resulted. But while the geologist can distinguish on general principles many cases of 'doubtful ground' it is also true that no existing geological maps are sufficiently detailed to show the slight variations in quality which are of the highest importance to the industry. Though a tract of land may be safely rated as gravel-bearing and the thickness known within limits, it needs to be proved in detail by boring or by trial-holes to locate failures and demonstrate the quality and its variation.

Another important limitation on choice of site in all dry workings is the availability of water for washing the gravel. The amounts used are comparatively small for the process is 'cyclic' and the main loss is by evaporation. Surface supplies from small streams sometimes suffice, but special borings for water are often necessary. These are subject to the limitations imposed by the Water Act of 1945.[1]

Transport

The greatest single factor in the existing localisation of the industry is accessibility to markets. This may sound the most palpable truism yet it has been persistently ignored in much ill-informed comment and in many planning proposals. But gravel is of low intrinsic value and the cost of transport is a major factor in the price of the material ; loading followed by a two-mile haul from the pit adds 30 per cent to the total cost of excavating and dressing a 'medium quality graded' gravel.[2] Each additional radial mile of haulage raises the cost by about sixpence a cubic yard and at a distance of thirteen miles from the pit the cost price is nearly doubled. These facts show the over-riding importance of 'accessibility'. Gravel is in fact more sensitive to the effect of transport charges than any other mineral. Not only is its 'pit-head' price low, but gravel characteristically moves in small loads to a wide variety of destinations which themselves are constantly changing as building and road con-struction shift from one place to another. In this respect it differs fundamentally from the Jurassic ironstones, for example, the pit-head price of which is similar to or even less than that of gravel, but which move in whole train-loads at regular inter-vals to fixed destinations which may be 150 miles or more from the orefield. The end-product too, usually concrete in some shape or form, is of relatively low value compared with the steel which forms the ultimate product of ironstone. These facts are responsible for the virtually complete dominance of road trans-port in the gravel industry. The life of a pit is seldom long enough to warrant the laying of special railway lines, as for a

---

[1] Section 14 of this Act empowers the Minister to designate areas in which the protection of water supplies is desirable, and in which new borings for industrial purposes will be subject to licence and, by implication, permitted only if the Minister is satisfied that existing waterworks will not be adversely affected. [2] *Report*, p. 29

new coal mine or an ironstone quarry, though a few large pits alongside existing lines may sometimes make use of special sidings, and the advantages of the lorry for distribution to the actual sites where the gravel is needed are obvious enough. Water transport, except for an occasional pit situated alongside a canal or navigable river, such as the Trent, plays a negligible part.

The question of substitutes for gravel is one that is often raised. It is impossible to predict what may happen in civil engineering during the next few decades, just as few could have foreseen, say in 1913, the extraordinary developments which have taken place since 1920. At present, crushed stone (generally sandstone, limestone, or igneous rock) is virtually the only competitor with gravel, and that only locally or in special circumstances, for the pit-head price of these other rocks is appreciably greater. In Cornwall and west Devon, counties which are not rich in ordinary gravel deposits, the sand and fine gravel from the waste heaps of the china-clay industry are in a competitive position.

It is vital that these economic considerations should be more widely understood. But the solution of any local problem requires more than the broad recognition of the principles ; a knowledge of the actual movement of gravel from pit to market, with the range of delivery points, is indispensable. The Advisory Committee on which we are serving has been able, with the co-operation of the industry, to trace the flow and destination of the product and to give definitive meaning to the concept of 'service areas', which will form an important part of the basis of its recommendations outside the Greater London area.

Land-use Competition

The foregoing paragraphs summarise the factors behind the geographical pattern of sand and gravel working and the problems inherent in that pattern. Since several forms of land-use compete for the same land we have, from one aspect, a strictly economic problem, the 'disposal of scarce means between alternative uses'. But, to isolate the geographical aspect of the problem and see it in correct perspective, it is essential to realise that it is one of *conservation* rather than short-term disposal.

The general argument for home-grown food is still unacceptable to economists. To Professor Jewkes, for instance, it appears as 'the oldest and most deeply embedded fallacy of economic thinking' that 'whatever the cost it is always a good thing for a country to produce its own food'.[1] In the same way, economists are unimpressed by claims based upon differential land quality. For Marshall[2] there was 'no absolute measure of the richness or fertility of land. Even if there be no change in the arts of production, a mere increase in the demand for produce may reverse the order in which two adjacent pieces of land rank as regards fertility.' But when all allowance is made for the element of truth in these assertions they clearly miss the point of our problem. On urban borders where the pressure on land is greatest, what it is desired to conserve is not food as such but a specific variety of perishable food in which freshness is at a premium. More generally, the need is to preserve a specific variety of land which, when with other varieties, is a necessary element in efficient husbandry. Many of the riverine gravel tracts, say those of the Trent valley, lie too far from towns for market-gardening ; but taken with the adjacent heavier lands, and above all, considered in their relation to the water-table, they are an integral part of a whole. Even if it were considered that, under the control of prices, urban development must always take precedence of rural land-use, it would in most regions still be necessary to attempt to preserve some balance between 'gravel' and 'non-gravel' lands used for farming. Statements such as Marshall's, true enough in their limited context, do not carry the corollary that land is a substance of which the variation can be ignored. The conversion of London Clay soils at whatever cost into market-garden soils, even remotely comparable in productivity with those of the brickearths, is not so much uneconomic, as it is physically impossible. One might argue that a change in taste or fashion, for grassland products at the expense of those of 'truck-farming', might reverse the order of the two types of land in terms of 'fertility'. But no-one reasonably conversant with agricultural economic history is likely to be impressed by this argument.

While there are importable surpluses of food, the restriction

[1] J. Jewkes, *Ordeal by Planning* (Macmillan, 1948), p. 235
[2] A. Marshall, *Principles of Economics* (Macmillan, 8th ed., 1938), p. 157

of urban land-use in this country at the expense of agriculture *is* in a sense uneconomic ; but this statement is governed so completely by the conditional clause that in the present state of world food supply it will carry little of comfort or guidance

Even if we concede that the margin of transfer between urban and rural land-use is, and should be, under the simple governance of prices, the problem of conservation remains a real one. That unregulated urban expansion permanently sterilises valuable mineral resources is evidently a new and unwelcome truth to town planners. It would argue a lamentable ignorance of that awareness of environment for which the geographer strives, a peculiarly unecological blunder, a veritable killing of the goose that lays the golden eggs, if towns occluded their own building material The material substance of a modern town is often no less the direct product of the 'soil' on which it stands than its brick-built nineteenth-century predecessor or the stone-built Cotswold village.

The problem is essentially one of conservation, for surface mineral working is a destructive activity unless some after-treatment of the site promises an eventual economic return. This is notably true in the extraction of sand and gravel, which consumes relatively more land because of the general thinness of the deposits worked ; and generally land of high quality, judged either by urban or rural values. This throws doubt on the efficiency of the price-mechanism to harmonise these competing forms of land use. Fifteen years' working of a fifty-acre site can leave behind a truly negative area, a sterile and offensive wilderness breeding mosquitoes and weeds. Such a situation is not beyond cure, but production must at least be conducted with a view to possible after-treatment and the choice of sites must be partly controlled by this. With 'dry workings', at least in land of the better agricultural class, it is normally possible to regain a considerable measure of productivity by simple levelling without refilling ; in some cases indeed the soil is improved by the removal of the gravel cover. The filling of the wet workings which so greatly disturb the public mind will also normally follow in areas within economic range of filling material, and will ensure the restoration of a surface suitable at least for light industry or normal residential buildings. These are the natural and appropriate forms of land-use near many urban borders, and 'dry' sites when levelled are not less suitable. The technical problems are those of 'soil mechanics', as it is inaptly termed by the engineers, not of pedology. The recovery of 'fertility' in the case of wet workings is a much more complicated problem and involves both filling and top-dressing. The circumstances are very different from those of opencast working for coal and iron. The virtually complete removal of the original rock substance requires the substitution of extraneous material, and in effect the construction of an artificial soil profile above it. The replacing of what is loosely known as top-soil on an extraneous filling is insufficient ; even if the filling is available and cheap it may be quite unsuitable as a 'soil parent', and may adversely transform conditions in respect of 'root-room', water properties, etc. Pending the institution of controlled experiment we are, and shall remain, ignorant of the possibilities and no solution can yet be offered for the wet pits beyond the economic range of 'urban' filling material.

The problem of conservation is further illustrated in the not infrequent choice between the exploitation of riverine gravel in agricultural utilisation and gravel underlying adjacent common land of low agricultural rating. We have seen that considerations of quality and ease of working generally favour the former, but granted a supply of water for washing the higher-level gravels this is not necessarily the case. The conflict is then between agriculture and amenity. Commoners' rights are also involved and special legislation is necessary to make the land available at all ; it is not, in the normal way, in the market. The simple price-mechanism cannot harmonise the conflict, if only because 'public goods' in the form of amenity are in question.

The starting-point of any investigation must be that the present production centres are where they are for good and essentially unalterable reasons, geological and economic, though not that the distribution is an optimum one or that a measure of rational redistribution cannot be envisaged. The geological facts are by no means obvious to surface inspection and many of them have been wholly or partly unfamiliar to the industry, which has not been well enough served by geological information to adapt itself intelligently to its environment. Areas that are not worked are not *ipso facto* uneconomic ; often they are simply unknown to the industry. Even so a more serious difficulty has been the non-availability of land. Some areas have been more tightly held than others, and in some the decision of a large landowner has completely excluded gravel working. Yet within these limitations the distribution of the industry is 'economic', and could not be radically or rapidly changed without a complete disruption of supply and an increase in price that would be reflected in every branch of building and road construction. The habit of planning consultants, and local interests generally, of saying in effect 'No doubt gravel must be worked somewhere, but not here' is destructive to all rational conservation and often rests on a grave factual error, that gravel reserves are vast and inexhaustible.[1] Having regard to quality and accessibility, the position is quite otherwise. A similar illusion might be sustained in regard to coal if all carbonaceous deposits, including the vast tracts of upland and lowland peat, were shown on one map and regarded as economic substitutes for one another. The differences within what are generically 'gravels' are hardly less. Building sand is no substitute for sharp concreting sand, and 'hoggin' or 'binding gravel' will not yield concrete aggregates without prohibitively expensive costs in washing.

A wholesale transplantation of the industry cannot be contemplated. Within each 'service area' it is necessary to find land for the maintenance of present supplies and to find and protect reserves for the future. Local substitutes such as crushed stone, or sand and gravel dredged off-shore, may relieve the pressure, though the latter source involves difficult problems of longshore protection. It is not possible to leave the problem at the mercy of the chance availability of land, which would be to invite further irreparable damage to agriculture and amenity, or alternatively to permit a serious and possibly complete interruption of supply.

The nature of the problem in the London area may be appreciated from the following figures. The largest source of supply, the western or west Middlesex region, furnishes more than 40 per cent of the total production and more than three times that of any other source. In 1939 there were here 23,000 acres of gravel land potentially workable. More than a quarter of this has since been sterilised by Heathrow airport, water reservoirs,

---

[1] See, for example, *Greater London Plan, 1944*, pp. 59, 86

and further building. Of the remainder, much is doubtful on geological grounds and almost as much is already ear-marked for urban development. The allocation of the remaining 11,000 acres presents a problem of great importance but little analytical interest. If agricultural reservations are deemed to be justified, nearly 5,000 acres can properly be rated as first-class land ; the area for excavation, the remaining 6,000 odd acres, represents a reserve which even at present rates of production would be exhausted within fifty years. The solution proposed involves progressively laying under contribution the gravels of the Vale of St Albans, of poorer quality but underlying less productive agricultural land. These gravels, it should be noted, are at about the same radial distance from London as those of west Middlesex. Here a reserve of some 9,000 acres may be regarded as potentially making good the deficiency, not only in west Middlesex, but in the other metropolitan areas, some of which will be exhausted even earlier.

Nowhere else in the country, save locally in the West Midlands, Yorkshire, and Lancashire, does past or impending urban development narrow the choice as sharply as this. In such regions as the Trent valley, Hampshire, or the middle and upper Thames, gravel working is of appreciable importance, but the areas involved, now or in any expectable future, form a relatively small part of the total gravel field. Yet, considered by individual service areas, the problem of land allocation remains sufficiently formidable. The gross areas of the gravel field must first be reduced by allowing for land that is geologically doubtful, or planned for other development. In the considerable remaining area, agricultural and mineral production must continue side by side, and the problem is, in effect, to divide the land between them. It is both impossible and undesirable to prescribe in detail the site of future gravel workings. Even on the quite untenable assumption that the required grades of product, modes of working, and general 'market patterns' were destined to remain unchanged, the available geological knowledge is insufficient for any such detailed plan. For considerable areas it is still true that no 'drift maps' are available. One cannot say 'Here and nowhere else should gravel be worked during the next x years', but it is not impossible to propose broad allocations of land compatible with rational conservation. For the reasons we have indicated it will nearly always be found that the optimum areas for agricultural and mineral production overlap, but they are not co-extensive and it is sometimes possible to reserve land for agriculture and find alternative sites for the gravel industry. Such a choice may justly claim the title of planning for conservation, but if it is to be successful it is essential that the areas allocated for gravel working should be much more extensive than the areas actually to be worked. Only so can the necessary flexibility of future development be safeguarded, and the indispensable but unformulated knowledge of the industry turned to account. A producer requiring a 'replacement pit' has definitive requirements in respect of character of product and mode of working. To satisfy these he must be offered, in effect, a range of sites. Of these he will pick only one ; the remainder will continue in agricultural production. It is vital that this principle should be grasped by all concerned with the administration of development plans.

There is a natural, and perhaps excusable tendency for agriculture to claim too much in such cases. One may reasonably reject the extreme claim of economists that the conservation of agricultural land is a fallacy, without ignoring its underlying element of truth. Agriculture is justly served by the reservation of the highest quality land, but much depends on the definition of 'best'. If this is deemed to include say two-thirds or five-sixths of the total gravel field, the residue is often demonstrably insufficient. If the areas for mineral working were too narrowly circumscribed, the last state of agriculture would, in our judgment, be worse than the first ; for emergency needs of mineral production, whether in peace or war, will then perforce invade the agricultural reservation, taking perhaps the best rather than the merely better land. The problem is misconceived if it is seen as a contest between agricultural and mineral reservations. What is required is that the industry should be given a measure of priority in chosen areas of wide extent. Such a policy avoids the criticism that it is in general 'restrictive'. It is restrictive as far as absolute reservations for agriculture are made, but may yet result in allowing the industry a wider and freer choice of land than it has customarily had in the past and in enabling it thus to adapt itself to technological progress and market vicissitudes, and to minimise its costs of production.

It is unlikely that such a policy would fail to find favour in all quarters, but its very possibility depends upon continued research in the field of land classification. Fully informed proposals on the lines indicated would require the completion of the geological drift survey and a rapid progress in soil survey in which this country has made only a limited beginning. Pending the completion of such surveys it is vital that other methods of land classification should be fully investigated and discussed, and this is a task to which geographers may properly commit themselves.

16

The Conservation of Natural Resources

OUR title announces a subject so illimitably large that time alone and, not less, the extent of my own ignorance make it impossible to cover the field. In its widest reference the subject embraces the whole substance and central problem of human geography properly conceived. Seen in the broadest terms, the fight for the survival of our species is now on, and it is concerned not only with modes of government and risks of war but with the material basis of civilisation itself—the problem of the earth as a habitable planet. This wider question would lead us far from what must be our more immediate concern now, the position of our own country. Yet since the problems of conservation in Britain are a special instance within the wider field, and indissolubly linked with it, we must begin by seeking some sort of world perspective.

World Resources : the Present Situation

The tale of history has run so fast within our lifetime that it requires an effort to realise how rapidly our picture of world resources has changed. Less than a quarter of a century ago it was widely, if uncritically, assumed that large exportable surpluses of food such as the nineteenth century knew were still in indefinite prospect, and that our mineral supplies and sources of power, though evidently not inexhaustible, were ample enough to put a premium upon waste. Earlier warnings had proved mistaken ; one recalls Crooke's forecast to the British Association in 1898 that agricultural expansion was nearing its end, and the earlier alarming conclusion of W. S. Jevons, in 1865, that British coal would be exhausted by 1970. The course of events falsified these predictions and all too readily there grew up a complacent attitude which rated any attempt at intelligent foresight at a discount. The foolish and misleading catch-phrase 'poverty in the midst of plenty' reflected in fact local deficiencies of purchasing power.

But it bred an unthinking complacency concerning the supposed unlimited endowment of our planet in food and raw materials.

It needs no words of mine to emphasise the entirely changed intellectual climate of the present time. Yet even now the ominous facts are not widely enough recognised and need constant repetition. With a world population rising by twenty millions annually—more than 50,000 a day—and with total cultivable area, even including the wet forests of the inter-tropical belt, estimated at less than a third of the total land surface, the prospects of adequate food supply appear dubious to many well qualified to judge. The words of Sir John Boyd Orr are already famous— 'the whole human race is rumbling on to destruction. There is only a fifty-fifty chance of getting over this food problem'.[1] In his posthumous book *The Estate of Man*, Michael Roberts thus concluded his objective and laconic survey of the problem of Food and People : 'There is indeed no answer but only a hope. A more careful survey of the world's resources may reveal that there is still a breathing space of a generation or so, during which the world's production of food can increase and during that time western ideas might so influence the East that even in India the population might be stabilised at something not too far above the present level.'[2] Professor C. B. Fawcett's estimates of population capacity have a slightly more optimistic tone since he concludes that, with our present powers of production, the world may be able to support three times its present population in reasonable comfort.[3] Yet he adds that at present rates of increase that number will be reached less than a century from now. The most recent authoritative pronouncement on the subject, by Sir John Russell, in his address to the British Association at Newcastle,[4] is also instinct with courage and hope, though he makes no forecast of possible food production.

The limit at any time is set by the efficiency of the plant as a transformer of radiant energy ; at present this does not exceed 5 per cent and, reckoned on the basis of the amount of food produced, it is much less. Whether this can ever be raised, whether we can ever do more than increase the proportion of assimilation products useful as food cannot be said. But the

[1] *The Times*, 5 May 1948
[2] M. Roberts, *The Estate of Man* (Faber, 1951), p. 26
[3] *The Advancement of Science* 4 (1947), 146–7
[4] *The Advancement of Science* 6 (1949), 132

present limitations to food production : utilisation of 7 to 10 per cent only of the earth's surface ; conversion by the animal of 10 to 25 per cent only of its food into human food ; and the fixation by the plant of no more than 5 per cent of the radiant energy received ; these are challenges to agricultural science which its workers are vigorously taking up.

Nor is it only in food that impending world shortages threaten. We are consuming forest products so fast that the present total forest area of the United States, including plantations, is only half the original area. Meanwhile a single issue of a New York Sunday newspaper costs twenty acres of timber. The known reserves of petroleum will be exhausted at the present rate of consumption in twenty-two years ; even if we make full allowance for new oil-field dicoveries, of which the geological possibility can be fairly closely assessed, natural petroleum can hardly outlast the present century. The best of the Carboniferous coals will also become scarce within a century, leaving, however, a large potential reserve of lower-grade fuels. Three years ago the British Association discussed the conservation of the very chemical elements themselves in the light of the fact that the exhaustion of the more concentrated mineral deposits must ensue in a matter of decades. It is claimed that more minerals have been raised in the United States since 1900 than from the whole world during the whole of previous history.[1]

Conservation Problems in Great Britain

It is in the light of such a general world picture that we have to consider the conservation problems of our own country. It will be wise at the outset to note, as a guiding principle, that our problem is not simply technological but economic. This no doubt is a trite observation but one regretfully concludes that it needs repeated emphasis. However important in the long run the application of science to technique may prove, vague non-quantitative aspirations for more home-grown food or bigger and cheaper exports have no meaning whatever unless brought to measurement by the methods of economics in its study of the 'disposal of scarce means between alternative uses'. For the principles which control such disposal, rationally conceived, are,

[1] *The Advancement of Science* 5 (1949), 338

in the best and truest sense of the word, economical and provide the only road to an intelligent policy of conservation.

Land-use

Let us turn first to the basic question of land-use, which underlies most of our conservation problems. The first half of the century has witnessed the loss of two million acres of agricultural land to urban expansion in its widest sense, surface mineral working, and the needs of the defence services. If we seek to forecast the future loss of the next twenty-five years we can call on a certain amount of detailed evidence, but since the data are incomplete and, where available, are still confidential, I prefer an estimate based on the trend of the first half of the century. This indicates that we shall see the withdrawal of a further million acres from agriculture, irrespective of the demands of the Service departments. Including the losses of earlier years this would mean that by 1975 there would be about six million acres of non-agricultural land in England and Wales, about 16 per cent of the total land area, and the amount of such land would have doubled during the present century. It may be best visualised by regarding it as approximately equal in area to Yorkshire, Lancashire, Durham, and Westmoreland taken together, or of Cornwall, Devon, Somerset, Dorset, Wiltshire, and Gloucestershire.

The danger of stating losses of agricultural land in such terms is that we may be tempted at once to contemplate a ridiculous and Canute-like policy of seeking to conserve agricultural land as such against all comers. The figures, of course, justify no such policy. In so far as we now require to increase home-grown food there is still ample room on land of good or potentially good quality to achieve, by intensification, all that we are likely to be able to afford. I cannot believe that a modern Cobbett crossing our island from, say, Plymouth, and travelling by road via Exeter, Taunton, Bristol, Gloucester, Leicester, Newark, and Yorkshire to Newcastle could form any other conclusion. The real significance of the large area of non-agricultural land is as a reminder to us of the vast investment of skill and capital represented in our industrial growth. A. C. Hobson [1] has recently reviewed post-war changes in the ' Great

[1] *Econ. Journ.* **61** (1951), 562–76

Industrial Belt ', which has been described as extending from Haslemere and Gravesend in the south to Colwyn Bay and Knaresborough in the north. Population and industry are still increasing in this zone, and we may conclude that it still affords the best location for most of our industries. But if this great industrial belt at once expresses the tendencies of past economic history and foreshadows things to come, there is ample room within it for profitable agricultural production. Our modern Cobbett, following the route I have described, would have crossed the belt from side to side but might, in fact, hardly have noticed it.

This view of the position is a necessary counterpoise to exaggerated conclusions and panic-stricken proposals which involve seeing and administering our problem as a contest of agriculture *versus* the rest. If conservation in any real sense of the word is to be sought it is fatal for any one industry or interest to be given over-all priority or final veto in land use. Yet this is what the less reflective advocates of agriculture persistently claim. When their cry is woe ! woe ! at each and every housing project or proposal for mineral working they need to be reminded of what our modern Cobbett might have seen in his journey through parts of the East Midlands. Here, not twenty miles from fair and famous fatting pastures, and on soils potentially of the same quality, he would find poor and ill-equipped farming on land calling urgently for draining, and requiring for its salvation steel, concrete, and bricks. It is only at the cost of using other land for such products that these degenerate and depressing acres can be brought to play their appropriate part in the provision of food.

I am necessarily conscious that these statements, though, as I believe, unarguably true and consistently urged by economists, will be unwelcome to the more violent, or less intelligent, contenders for the agricultural interest. But, in fact, the powerful case that can be made for a rational land conservation policy is only confused and weakened by some of the almost inspired muddle-headedness stridently offered in its support. Though the state of the world market is disastrously changed and concern for home-grown food is appropriate, the arguments of a former day must remain in our minds. Whatever expedients present conditions may justify we must be prepared to find that subsidised agriculture keeps out cheaper food and employs labour and capital which could be better used otherwise, and with the ultimate production of more food. We cannot expand home food production as a thing in itself without effect upon manufacture for export. There is a case for seeking to change, within limits, the amounts in which our scarce resources of lands are disposed as between agricultural and non-agricultural use, but none for pretending to forget that 'more of this means less of that'.

All this is evidently not to say, however, that there is no case for general economy in the use of land. It is right that we should reconsider the appropriate housing densities of new development and critically examine the dangerous assumption that garden cultivation more than makes good the loss of agricultural production. It is reasonable that we should re-examine our over-lavish projects for the provision of playing-fields. It is essential that we should show sufficient foresight to facilitate the prior extraction of surface minerals when such is consistent with planned after-use, and also to secure some profitable use of the vast deserts controlled, but only very occasionally used, by the Service departments. All these are questions of general land economy. Much more interesting in analysis and difficult in practice are the problems of securing the best use of specific sorts of land.

Here has seemed to lie the gravamen of the case for land conservation. We find, however, two utterly contrasted approaches to the matter. On the one hand is the view which economists still tend to hold that land quality is largely man-made. They tacitly recognise some variations in inherent quality but show little awareness of how important these may be. It is true, of course, that in long-settled country, such as the British lowland areas, the general uniformity of the agricultural landscape all too readily conceals its geological basis from the inexpert eye. Vast totals of labour and capital have upgraded many intrinsically poorer soils and blurred the lines of what might be regarded as the natural pattern of land-use. Closer study shows that this impression is misleading. We must not lose sight of the fact that no conceivable expenditure of resources could replace the loss of some types of land by the conversion of others. To suppose the contrary is almost as scientifically ridiculous as to contemplate the conversion of chalk to cheese by a complex and expensive process of molecular replacement. Nature sets limits to the convertibility of soils. This is obvious enough with land in extreme climatic or physiographic situations ; it seems to be apprehended with difficulty within the limits of the wide land-class that may be vaguely rated as fertile.

It is important to examine the reasons in this misunderstanding. They arise primarily because it is impossible to express grades of soil quality in precise quantitative terms. Neither rent nor the annual value of produce will avail to express differences as seen by the soil conservator concerned with the land as a long-term asset. In southwest Sussex, for example, the rent of brickearth land is nearly twice that of neighbouring gravel land. More generally, it would be true to say that, at 1950 prices, a wide range of farming types in England and Wales had an annual value of output per acre of between £25 and £30. Fruit, hop-growing, and market-growing lands gave values up to three times as high. But whether one thinks of the higher rent of the physically superior soil or of value of output, the community value, in the fullest sense, of the better land rests on the fact that it is more adaptable—adaptable at minimum cost to almost any form of agriculture production which world vicissitudes or national interest might require. It is to this quality, deriving essentially from geological parentage, that it is impossible to attach a calculable monetary value in the short term. The very best of our soils are essentially applicable to the production of high-grade cereal crops, root crops for animal fodder, or to either milk or horticultural products. Our power to maintain maximum flexibility in production depends upon the deliberate conservation of such soils.

Another vital principle readily forgotten is the dependence of profitable farming systems on associated land types complementary to one another. The alienation of valley floor lands can throw much larger acreages of hill land out of production. Local resources of arable soils, not intrinsically of very high quality, may form an indispensable element in an area dominated by heavy or low-lying pasture lands.

Land Allocation

In all this I am merely repeating what has been often and forcibly urged by others, notably by my colleague Professor Dudley Stamp. Such contribution as I can make derives from five years' service on the Waters Committee [1] and the opportunity thus afforded of watching and participating in practical attempts at land allocation. I should be entirely lacking in frankness if I did not admit the difficulty of such work. When the problem is broken down into local areas one is impressed not by the general total of land alienated from agriculture but the relatively small amount and slow rate of urban growth and related forms of encroachment. In a local context, moreover, no general system of land classification affords much help ; each case is a special case dependent on the local association of land types and the fact that each farm is itself a unit. It is, no doubt, unarguably true that there is an optimum use for each acre of land in this small country, but it is an unwarrantable assumption that such use is revealed to simple inspection or even to careful reflection. 'In economic action,' writes L. M. Lachman,[2] 'the problem to be solved is to devise a plan for the allocation of scarce resources to alternative wants in such a way as to maximise satisfaction . . . and since the elements of the plan are quantifiable, if not measurable, the problem and its solution can be illustrated *more mathematico*. However, that a problem has a determinate solution does not entail that those attempting its solution will actually succeed. Otherwise there would be no failures in examinations or in business.' In the case before us it is doubtful if the elements are even quantifiable ; small wonder if the solution proves indeterminate. In reviewing the Scott report, D. H. Macgregor [3] prophetically imagined 'a case of some ten acres leading to a hectic discussion in the departments' and thought it would be interesting to see the papers in which the contestants extended themselves and each other. His prophecy has been exactly fulfilled, though it is with consternation rather than interest that one observes the resulting inroads on our scarce resources of paper.

This is not to say, however, that the effort to save land by the guidance or direction of development is wasted effort. We have struggled with this problem, in any definitive sense, for less than a decade and we inevitably learn from our mistakes. The case for the conservation of the better land rests on the manifest fact

---

[1] *Report of the Advisory Committee on Sand and Gravel*, Parts 1 and 2 (H.M.S.O., 1948) and succeeding parts still in course of publication

[2] *Economica* 10, No. 37 (1943), 15   [3] *Economica* 10, No. 39 (1943), 11

that past development has often been fortuitous and has involved avoidable waste.   If a salient example were needed one has only to call in evidence the Thames valley plain west of London. If our present powers had been available only twenty years ago, and, what is equally important, our present modes of thinking had been then prevalent, large areas of some of the best agricultural land in the world, not to mention enormous reserves of building materials, could have been saved without any despite whatever to housing or industrial development.   It is naïve in the extreme to conclude that the 'invisible hand' not only maximises present satisfactions but rationally conserves limited and irreplaceable resources for the future.   The sorry story of soil erosion over the greater part of the world constitutes a sufficient rebuttal of this peculiar assumption.   It has been well said by Miss Barbara Wootton [1] that a fanatical belief in 'marked norms' involves the corollary that 'all decisions which are consciously and deliberately taken by a social group are the decisions which are almost certain to be wrong'.   Wrong in part they may certainly be, but at their worst they are hardly worse than the integrated results of individual ignorance and short-sightedness, and at their best they are incomparably better.   Not the least of the benefits of our attempt to plan land-use is that it tends, if done openly, to inculcate that awareness of environment for which the geographer strives, but of which so many of his associates fail to show evidence.

After-Treatment of Worked-out Excavations

Let us turn now to another aspect of land conservation, the after-treatment of the worked-out excavations for surface minerals. With opencast coal working reinstatement of the surface normally follows the operation, and similar provision, involving a levy on production, has recently been made for ironstone workings. The working at present or projected rates of the three other major categories of surface minerals—sand and gravel, brick-clays, and chalk and clay for cement—requires the consumption of about 2,250 acres of land annually, of which 80 per cent. (*c.* 1,800 acres) is accountable to sand and gravel. About two-thirds of the latter production involves excavation below the water-table in wet pits, which survive as deep lagoons.

[1] B. Wootton, *Lament for Economics* (Allen & Unwin, 1933), p. 174

The plea that all such excavations should be filled or adapted to some beneficial after-use makes a ready and evident appeal, but the problem is one of great difficulty. It is essential to realise that pits for sand and gravel, brick-clays, or cement materials differ entirely from those for opencast coal or iron-stone in respect of the availability of filling materials. In the latter, a relatively thin seam is extracted from amongst associated strata, which remain as overburden. In the former the greater part of the deposit as a whole is removed ; overburden is at best small in amount and often negligible. Nevertheless dry workings present no insuperable problems for after-treatment ; the levelling of pit floors and the battening of sides will normally leave a surface available for cultivation or for building, or, if the pit bites into a sharp hillside, for normal re-vegetation. With wet pits, unless it is desired to preserve them as landscaped lagoons, two suc-cessive problems arise. the filling of the pit and, if cultivation is in view, top-dressing of the fill.

It would be improper, even if it were possible, for me to seek to forecast the findings of the Waters Committee on this vexed problem. But there are certain relevant facts to which attention may be drawn. On urban borders generally, but more especially around London, many wet pits have been, and are being, filled profitably for the operators since they serve as tips for builders' débris and other inactive urban rubbish. Locally at least, the condition is almost that of too much rubbish chasing too few holes, and the perimeter of economic filling has demonstrably moved outwards from London in response to this pressure. The problem can be seen in its proper perspective only by considering its regional incidence. The Greater London area is answerable for nearly a third of the production of wet pits If we include with it the middle and upper Thames area, Kent, Sussex, and Hampshire, all regions of similar geological consitution, the wet acreage makes up more than half of the national total. Only in three other areas is the wet acreage large, the Bedfordshire area and the Fenland margins, parts of the West Riding of Yorkshire, and the Trent valley. These together account for rather over two-fifths of the total. If we add this quota to that of the south-eastern regions, 85 per cent of the total wet acreage is accounted for.

The general scarcity of filling material and the high cost of

A. W. Ashby that the mechanisation of agriculture during this century has saved three million acres for human food production in the place of land formerly used for the support of horses. Despite our present difficulties of coal production we have reserves conservatively estimated as sufficient for at least 500 years—while the estimated reserves of iron ore would avail at present rates for 150 years. The mineral base of our economy remains, both absolutely and relatively, very strong indeed.

Our greatest need is for the deliberate, patient, and objective study of the over-all picture of resources and their disposal. There, as I have indicated, is seen the rôle of the modern geographer in the field of what has been called geo-technics. I must confess to some bewilderment, not untinged at times by exasperation, when I meet the recurrent doubt as to what geographers do or can do in this field. The modern administrator seeks to construct his over-all picture, and maintain his characteristic balance of view, and he is accustomed to deal with blinkered technical specialists, competent within their own fields, who are summoned as fact-providers but not expected to have views on policy. In this respect your geographer is no doubt a dangerous man since his training in the technique and discipline of being knowledge-ably aware of environment moves him inevitably to form opinions on policy. What is true in the more general field of planning is, in any case, conspicuously true in the field of town planning. I hope to see this Institute increasingly aware of the geographical implications of its own special interests, and effectively convinced that the planning of the towns is but a single, if important, aspect of the general problem of the economical disposal of resources.

Conclusion

THE title of this book will appear to many to embody an implicit question, almost perhaps a challenge. It is common though not universal in British universities to find geography ranked both as a Science and an Arts subject. This dual rôle it sometimes shares rather curiously with mathematics and psychology ! But its admission as a Science subject in any strict or meaningful sense has often been opposed and would be far from generally conceded.

In the penultimate section of his major work on *The Nature of Geography*, Richard Hartshorne asks the question, 'What kind of a science is Geography?' In so speaking, he evidently begs the question, and at the same time implies some classification of the sciences. The chief division of the field which he finds significant and relevant to his discussion is that between the *nomothetic* or law-making sciences and those which are *idiographic*, i.e. concerned with unique phenomena. He rightly emphasises, however, that these are really two aspects of scientific knowledge, present, though in varying degree, in all branches of science. Similarly Professor F. A. Hayek [1] writes, 'Today it is almost heresy to suggest that scientific knowledge is not the sum of all knowledge', but notes at once that 'there is beyond question a body of very important knowledge which cannot possibly be called scientific in the sense of knowledge of general rules'. Rather similarly John Stuart Mill [2] in his *Logic* wrote that 'knowledge insufficient for prediction may be most valuable for guidance'. If we inquire how far geography ought and can hope to graduate from the idiographic to the nomothetic class, it is difficult to add usefully to Hartshorne's own discussion with which the present writer finds himself in substantial agreement. It is particularly important, in any case, to resist being simply bluffed by the characteristic *hubris* of modern physical or natural science into seeking to make geography what it has never been, is not, and cannot from its nature hope to be. It best illumines our problem

---

[1] *Individualism and the Economic Order* (Routledge, 1949), p. 80
[2] J. S. Mill, *Logic*, Book VI, chap. vi, p. 462

not to insist on the distinction between nomothetic and idiographic sciences, but to follow J. R. Baker [1] in recognising 'that
the whole of science may be cut right across and separated into
only two divisions, 'General Analytical Science' and Natural
History. The former is concerned with general rather than
particular facts and dependent in high degree upon experiment.
The basic sciences of the general analytical group are evidently
physics and chemistry, though genetics and physiology in their
modern guise are almost as fitly included in the group. By
contrast, Natural History covers the investigation of particular
objects and phenomena *which can be studied as wholes* mainly by
observation. Alternatively but with rather different implications,
Baker [2] cites the philosopher Thomas Hobbes who defined Natural
History 'as the history of such facts or effects of nature as have
no dependence on man's will, such as are the histories of metals,
plants, animals, *regions* and the like'. In view of the importance
of the regional concept in geography, it is of particular interest
to note the inclusion by Hobbes of 'regions' in this group. It
leads us to recall that as a student at Magdalen Hall, Oxford,
he took, we are told, little interest in scholastic philosophy and
devoted his leisure to books of travel and the study of maps and
charts. Though his later interests took other ground, here it
would seem was a potential founder of geography. He turned
in fact to geometry, physical science, and political philosophy.

Yet it is at once manifest that the geographer cannot and
will not accept a field from which facts and effects dependent
'on man's will' have no place. Regions are essentially human
creations, as the geographer sees and studies them. Here indeed
is the essential crux of the difficulty of ranging geography among
the sciences. Yet as Mackinder wrote,[3] 'we insist on the teaching and the grasping of geography as a whole'. The geographer
takes up 'the central position, looking equally on such parts
of science and such parts of history as are pertinent to his
enquiry'.

Or we may prefer C. O. Sauer's [4] alternative statement of

---

[1] *The Scientific Life* (Allen and Unwin, 1942), p. 92

[2] Baker, *op. cit.*, p. 92. The citation is taken from T. H. Huxley, 'On
the study of Biology' *Nature* 15 (1877), 219

[3] H. J. Mackinder, *Geog. Journ.* 9 (1887), 145

[4] Bowman Memorial Lectures (New York, 1952), p. 1

the position : 'Geography must take the risks of interpreting the meeting of natural and cultural history'. We need not deny that there are risks in attempting this interpretation, but it is fair to claim that the results so far achieved amply justify the risk.

It is nevertheless necessary to add that the problem at issue cannot be resolved on the basis of the work of any single geographer, as here offered. Such work may, it is hoped, illuminate the problem and even point ways to a solution ; it cannot do more. There are important aspects or divisions of the field, quite un-represented in the foregoing collection of papers, and some of them are at least equally, if not more important, in any balanced assessment of the subject. In Sauer's lectures cited above he includes 'a personal note' referring to his early observations in the upper Illinois valley and the questions then raised in his mind which have guided his further work and interests. Since this is essentially a personal book, it is perhaps permissible to make a similar brief reference. The writer owed his introduction to geography at school to one of Herbertson's students from Oxford. Accordingly he inherited that staple of British geo-graphical teaching, the system of 'Natural Regions'. Since, however, this was before the days of effective field work in schools, he was never shown a 'regional boundary'. In personal explora-tion, in seeking that awareness of environment which is surely the root of geographical thinking, he, in fact, discovered such a boundary on the northern margin of the London Basin. Look-ing northwards from Shenley or Essendon in Hertfordshire one is confronted by a manifest, almost dramatic, change in the landscape expressed both in land-forms and land-utilisation. Here indeed is a piece of 'areal differentiation'. In terms of Fig. 13 one is standing at these places on the northern margin of the south Hertfordshire plateau, overlooking the 'Vale of St Albans'. The latter has many of the characteristics of a French 'pays'.

It would be foolish and conceited to try to demonstrate too literal a consistency in the work of later years, yet it is true to say that the essential stimulus followed in those years came from such contrasts in the home countryside. Chesterton's dictum that 'to make a thing real you must make it local' proved true in the case of the present writer. A great part of the work recorded in the foregoing pages arose from interests thus awakened and consisted essentially in inquiry into the causes of the 'areal differences'. My friend and collaborator, Professor W. G. East, rightly detects my emphasis upon *a genetic approach*. To understand the contrast referred to above it is necessary to inquire how it has come about, in terms both of physical and human history. Hartshorne concedes, perhaps rather grudgingly, that 'the interpretation of present geographic features requires some knowledge of their historical development'. The purport of much here written goes further than that, and might be summarised by saying that no real or satisfying comprehension of 'areal difference' can be attained in any other way than by the study of historical development. But there is broad ground for disputation here, particularly if 'historical' is conceived as including geological history. Into this discussion we cannot here enter anew (cf. pages 89–93). It must suffice to say here briefly if rather bluntly that it appears foolish and futile to ignore the history or origin of land-forms in favour of their so-called 'functional significance'. The main reason for rejecting this attitude as we have already indicated is that it is intellectually despicable. It may be logical, but it is not psychological. If geography is to continue to attract young people of ability and insight it is essential that the subject should be exciting, not merely fitted into a logically coherent place in the general scheme of knowledge. No better example can be found of a vigorously exciting thesis, and one at the same time essentially genetic in approach, than Sauer's lectures (cited above) on 'Agricultural Origins and Dispersals'. The subject may seem to lead beyond the boundaries of geography in the narrower sense, but it is not in terms of such 'narrower sense' that the subject will be fostered and advanced. In the same way geomorphology remains an integral part of geography not by reason of mere convenience or historical accident, but because as developed by its great founder, W. M. Davis, it is intellectually exciting and satisfying. There has been some talk of what may be termed a purely 'geographical geomorphology', designed simply to serve the ends of the 'study of areal differentiation'. A singularly limited subject it would be in such terms and one with which the present writer wishes to have nothing to do.

The title of this book is not 'Geography as a Science' but 'The

Geographer as Scientist'. If we may revert finally to J. R. Baker's [1] discussion we may note his statement: 'It is rather because I believe there are two main kinds of scientist than because there are necessarily two main kinds of science that the division of science into two is suggested. Those who may be called the natural historians derive their impetus to research, I believe, directly from their ancestors' interest in the environment.' So undoubtedly it is in working practice. It is surely beyond doubt or argument that people of scientific training can enter the field of geography to its very great profit. What they will be unwise to attempt is to transform the subject into a piece of general analytical science, least of all by the exclusion of its wide contacts with human history and the social sciences. The writer finds it heartening to read the statement of a former colleague at the London School of Economics—Professor R. B. Smellie: 'It is one of the glories of modern knowledge that the study of geography has been transformed. . . . The proper study of man's habitat is one of the triumphs of the rational spirit.' This is the verdict of a professor of Political Science. It is true that when I once quoted this statement to the Herbertson Society at Oxford I was at once reminded that it occurs in a discourse entitled 'Why we read History', but the tribute is no less welcome on that account. If modern geography can be claimed as one of the triumphs of the rational spirit, geographers are unlikely to be greatly troubled by fears that their subject is not a 'science' in some special and probably narrow form. They might indeed accept as a statement of their aim the following words from Bacon's *New Atlantis* and claim that they 'maintain a trade—not for any commodity of matter but only for God's first creation which was Light—to have light I say of *the growth* of all parts of the world'.

[1] J. R. Baker, *op. cit.*, p. 94

Printed in Great Britain by
Thomas Nelson and Sons Ltd, Edinburgh